DOUGLAS AD/A-1 SKYRAIDER PART TWO,

Skyraider Part Two covers the active duty US Navy AD/A-1 squadrons. Because of the extremely prolific usage and sheer numbers of squadrons concerned, I did not have room to include the Naval and Marine reserve squadrons. Those squadrons will have to follow in Part Three and be included with Air Force squadrons or foreign aircraft as well as a discussion of Skyraider modeling. Part Three will not follow this volume directly and may not be addressed for a year or two.

Naval Fighters Number 98,

Douglas AD/A-1 Skyraider Part O covered the aircraft's developme, testing, variants, test units, research and development units, CAG units, CAG composite squadrons (VC), FASRONs, training squadrons, hacks (base / carrier / air group / squadron), VR-22 (COD squadron), lighter than air (LTA) squadrons and Marine active duty squadrons.

One of the training squadrons, Carrier Qualification Unit Four (CQTU-4), was left out of Part One and is included in Part Two on page 23. Many Skyraiders were assigned

either for special projects, or as an interim aircraft for a couple weeks to a month or two. These aircraft were not painted with new tail codes so cannot be identified. Some examples are: VF-131, VF-171 and VF-172.

Please note, most all squadron AD/A-1s had their prop hubs painted in the squadron colors. So if the tail color was orange so was the hub. This fact is rarely noted in the captions.

ATTACK SQUADRON ONE L, VA-1L

VA-1L XBT2D-1 BuNo 09100 taxiing at NAS Atlantic City, NJ, in 1948. (Roger Besecker)

Torpedo Squadron Fifty-Eight, VT-58, was established on 19 March 1946 and was assigned F6F-5Ns, SB2C-5s, and TBM-3/3Es. The squadron's mission was to work with the Commander Operational Development Force (COMOPDEVFOR) in the operational testing and evaluation of new weapons, equipment and methods for use by the fleet. From the results, training procedures, operating procedures and tactical doctrine were then recommended to the fleet. CVLG-58 was redesignated CVL-1G on 15 November 1946 and VT-58 was in turn redesignated VA-1L. On 13 December 1946, VF-1L was

established and VA-1L's F6F-5Ns were transferred to VF-1L. Also in December 1946, the squadron received TBM-3Ns and then on 4 March 1947 VA-1L added a fourth type of TBM to the squadron, the TBM-3W. CVLG-1 was assigned to the USS Saipan (CVL-48) and VF-1L and VA-1L were deployed aboard CVL-48 for its shakedown cruise to the Caribbean from 4 April to 5 May 1947. The squadron participated in a two week cruise to South America from 7 February through 24 February 1948. During this short deployment the squadron conducted a fly-over during the inauguration ceremonies in

Caracas, Venezuela, for President-elect Romulo Gallegos before returning to Atlantic City, NJ. In April 1948, an XBT2D-1 was received for evaluation and development of tactics. Then, in September 1948, CVLG-1 was disestablished and its assets and aircraft were absorbed by Aircraft Development Squadron Three (VX-3). At the time, the squadron was equipped with 2 XBT2D-1s, 8 TBM-3Es and 1 SNJ-4.

UTILITY SQUADRON ONE, VU-1 "UNIQUE ANTIQUERS"

Utility Squadron One (VU-1) was re-established at NAS Barbers Point, TH, on 20 July 1951, from the VU-7 detachment already stationed there. The squadron was originally equipped with five JD-1s (A-26s), five TBM-3Us, and one SNB-2P. The F9F Panther arrived in 1953 and the F9F Cougar followed shortly thereafter. A couple photo Banshees were also in use in 1959-60. On 1 September 1960, Guided Missile Group One (GMGRU-1) was disestablished and

Below, VU-1 target tug AD-5 BuNo 133878 at NAS Barbers Point, TH, in 1962. Aircraft was engine grey with yellow wings and horizontal tail surfaces and da-glo red tail and wing stripe. (Ginter collection) Bottom, VU-1 AD-5 BuNo 133898 on 17 October 1961. (USN) At right, VU-1 AD-5 target tug with target tow gear housing under the centerline at NAS Barbers Point, HI, on 20 June 1961. (USN) Bottom right, VU-1 A-1E BuNo 132449 at NAS Quonset Pt., RI, in 1964 where it awaits refurbishment and transfer to USAF stocks. (Ira Ward via Jim Sullivan)

its assets, including the FJ-3Ds, were transferred to VU-1.

VU-1 received its first Skyraiders, two AD-5s, in April 1959. Three to four were on hand through November 1960 and five to six from December 1960 through July 1962. The numbers went back to three to four AD-5/A-1Es from August through June 1963 with the last A-1E being retired in August 1963. The squadron provided fleet utility services until it was redesignated Fleet Composite Squadron One (VC-1) on 1 July 1965.

AIR DEVELOPMENT SQUADRON ONE, VX-1 NAS KEY WEST, FL

Air Development Squadron One (VX1) at NAS Key West, FL, was responsible for testing and development of anti-submarine and maritime patrol systems. To prosecute this mission, a number of Skyraiders were utilized, including three unique ASW versions, the AD-3S BuNos 122910-122911, AD-3E BuNos 122906-122907 and AD-5S BuNo 132470.

The two AD-3E sub hunters were created from two AD-3Ws and the two AD-3S sub killers were created from AD-3N stocks. The four aircraft allowed for two Hunter-Killer teams to be tested at VX-1. Essentially, the airframes remained unchanged, with only the planes' electronic guts being exchanged out. One AD-3S was equipped with an upgraded AN/APS-31 radar under the port wing and tested to ascertain if the single-seater could combine search and attack functions successfully in one aircraft. Instead, the Navy installed gear into the first batch of AD-4Ws and AD-4Ns so that they could perform the two air-

Above, VX-1 AD-4N BuNo 124150 armed with a yellow tipped ASW torpedo. (Paul Minert collection) Below, VX-1 AD-1Q BuNo 09366 over Boca Chica on 6 January 1949. (National Archives via Jim Sullivan)

craft Hunter-Killer mission. The one-off AD-5S had a crew of three and incorporated the Hunter-Killer mission in one airframe (see Skyraider Part One, pages 129-131).

Above, VX-1 AD-3S BuNo 122910 near Key West on 30 January 1950. (USN) Below, VX-1 AD-5S BuNo 132470 with MAD boom extended. (Paul Minert collection) Bottom, VX-1 AD-3E BuNo 122906 in flight in 1952. (USN)

ATTACK SQUADRON ONE B, VA-1B
ATTACK SQUADRON TWENTY - FOUR, VA-24 "BOMB-A-TOMS"

VA-24 was originally established with Curtiss Helldivers as Bombing Squadron Seventy-Four (VB-74) on 1 May 1945. VB-74 was redesignated Attack Squadron One B (VA-1B) on 15 November 1946 at NAAS Oceana, VA, where it received its first eleven AD-1 Skyraiders in August 1947. VA-1B was assigned to CVBG-1 aboard the USS Midway (CVB-41). VA-1B deployed to the Mediterranean on Midway from 29 October 1947 to 11 March 1948 with twenty-four AD-1s. Once in the Med, one aircraft was lost when LTJG Robert Reeb had a complete power failure after takeoff from Malta. No other incidents occurred at sea, but another Skyraider was forced down due to engine troubles on 10 August 1948. The following month, on 1 September 1948, VA-1B was redesignated Attack Squadron Twenty-Four (VA-24) and CVBG-1 became CVG-2. VA-24 was reassigned to the USS Coral Sea (CVB-43). While preparing for deployment the squadron was scheduled to upgrade to the AD-2, but a series of

Above, VA-1B AD-1 M/308 on deck on the Midway in 1947-48 while on deployment to the Med. (USN)

groundings due to engine problems prevented this. Instead, VA-24 transitioned to the F4U-4 Corsair in just nine days in late January/early February 1949 and deployed to the Med in March aboard CVB-43.

AIR DEVELOPMENT SQUADRON TWO, VX-2

NAAS Chincoteague, VA, was established on 5 March 1943 and in January 1946, the Naval Air Ordnance Test Station (NAOTS) was established there. VX-2 was then established to aid in the development and testing of air-to-air weapons and operated F6Fs, F7Fs, F8Fs, PB4Y-2s, PBYs, TBMs, JDs, and Culvers as well as a couple of Skyraiders. VX-2's tail code was "XD".

The first Skyraider received at VX-2 was an AD-3N in August 1954. It was replaced with an AD-3W in September which was flown through August 1955.

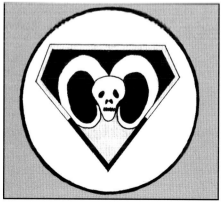

GUIDED MISSILE GROUP TWO, GMGRU-2

On 23 September 1955, VU-4 RAM Det and GMU-53 combined to form GMGRU-2 at Chincoteague, VA. It was responsible for the Atlantic Fleet Regulus I launches and RAM training. The squadron operated a number of Lockheed TV-2D (T-33) and FJ-3D drone control aircraft and deployed Dets to a number of carriers through 1957. GMGRU-2 also operat-

ed one AD-5 (tail code "DB") at Chincoteague from September 1956 to January 1958.

GMGRU-2 was redesignated GMSRON-2 on 1 July 1958. Missile launch restorability was given to GMU-51 while GMSRON-2 continued to provide missile recovery and drone services.

ATTACK SQUADRON TWO B, VA-2B
THE FIRST ATTACK SQUADRON TWENTY - FIVE, VA-25 "TIGERS"
THE SECOND ATTACK SQUADRON SIXTY - FIVE, VA-65 "TIGERS"

VA-2B was originally established as Torpedo Squadron Seventy-Four (VT-74) on 1 May 1945. A sister squadron of VB-74, it too was initially equipped with Curtiss Helldivers. Both squadrons were established to specifically operate from the new large deck USS Midway (CVB-41).

In August 1947, fourteen AD-1s were received to start replacing the squadron's SBW-5s and TBM-3s. Like VA-1B, VA-2B had twenty-four AD-1s on hand when they deployed to the Mediterranean aboard Midway from 29 October 1947 to 11 March 1948. For this deployment, VA-2B was assigned the 400 series aircraft numbers and VA-1B the 300 series. During the cruise, ports-of-call were: Gibraltar, Tangiers, Algeria, Malta, Genoa, Naples, Capri, Pompeii, Rome, Sicily, Taranto, Paris, and the Riviera. After its return to CONUS, VA-2B participated in Operation CAMID III in July and August 1948. During this close air support coverage of the operation's amphibious landings, VA-2B became the first Atlantic Fleet squadron to fire live Tiny Tim rockets.

On 1 September 1948, VA-2B was redesignated Attack Squadron Twenty-Five (VA-25) and CVBG-1 became CVG-2. The squadron's side

VA-25   VA-65

Below, VA-2B AD-1Q M-421 at NAAS Saufley Field, FL, in 1948. (Tailhook)

numbers were changed from 400 to 500 series and in September and October 1948, the squadron operated from NAF Barin Field, FL.

VA-25 and CVG-2 were reassigned to the USS Coral Sea (CVB-

Below, three VA-2B AD-1s in trail prepare to deck launch from Midway in late 1947. The upper portion of the tail was yellow and each aircraft is armed with four 5" HVAR rockets and carry an APS-4 radar pod under the left wing. M/424 is in the foreground followed by M/407. (USN)

43). Carrier work-ups for deployment began in March 1949, with the squadron sailing to the Med on 3 May 1949. They departed with nineteen AD-1s and two TBM-3Es and returned with eighteen AD-1s and the two TBM-3Es on 25 September 1949. Ports-of-call were: Cuba, Malta, Istanbul, Augusta, Palermo, Sicily, Naples, Leghorn, Rapatto, Cannes, Souda Bay, Crete, and Gibraltar.

After return to its home base at NAS Oceana, VA, VA-25 began its transition to the much improved AD-4 in November 1949, when three AD-4s were received. By the end of January 1950, the unit had transferred out all

Above, VA-2B/VA-25 AD-1 at NAAS Saufley Field, FL, in Aug.-Oct. 1948 during training at NAF Barin Field, AL. Markings included a yellow upper tail and a green gunnery pennant with a red dot on the nose. (NMNA) Below, VA-25 AD-1 BuNo 09141 aboard the USS Coral Sea (CVB-43) in April 1949. (USN)

of its AD-1s and had on-hand seventeen AD-4s.

On 1 August 1950, the squadron was reassigned to CVG-6 ("C" tail code) and the USS Franklin D. Roosevelt (CVB-42). VA-25 took its seventeen AD-4s aboard the FDR in October 1950 for carrier qualifications and the Air Group was transferred to NAF Weeksville, NC, in November 1950. The squadron's seventeen AD-4s along with an AD-4Q deployed to the Med aboard the FDR from 10 January to 18 May 1951. On return to

CONUS, the squadron relocated to CGAS Elizabeth City, NC, until it redeployed to the Med once more aboard Midway from 9 January to 5 May 1952. For this deployment, the squadron was flying a mixed bag of AD-4s and AD-4Ls. Ports-of-call were Oran, Augusta, Naples, Piraeus, Malta, Palermo, Cannes, Livorno, Taranto, and Gibraltar. Three months later the Midway and VA-25 deployed again, this time to the North Atlantic. For the cruise, which lasted from 26 August to 8 October 1952, the squadron fielded five AD-4Ls and

Above, VA-25 AD-1 (M/514) crashed on deck in late 1948. It was basically intact except for sheading its engine, and was returned to service. (USN) Bottom, VA-25 AD-4Q BuNo 124045 with twelve 5" HVARs and three 2,000 lb bombs on 18 May 1950. Prop hub and fin tip were green. (USN)

eleven AD-4s. Upon return to CONUS, the squadron was relocated to NAS Oceana, VA. One more Mediterranean deployment with the

Above, VA-25 AD-6 BuNo 134488 deck launches from the USS Coral Sea (CVA-43) in July 1954. Fin cap was green. (Jim Sullivan collection) Below, VA-25 AD-4 BuNo 128933 lands aboard the USS Midway in early 1952. Fin tip was green. (Tailhook)

Above, VA-65 AD-6 BuNo 137519 aboard CVA-11 on 10 January 1961. (USN) At right, orange-tailed VA-65 AD-6 BuNo 135322 on the deck of the Intrepid. (Paul Minert collection) Below, VA-65 AD-6 BuNo 139619 at MCAAS Yuma, AZ, on 3 December 1959 for the Naval Air Weapons Meet. Rudder trim was orange. Lightning bolt on fuselage was yellow. (William Swisher)

AD-4 was made aboard Midway from 1 December 1952 to 19 May 1953. After that the squadron transitioned to the AD-6 Skyraider, receiving ten in February 1954 while aboard the Midway in the Med. Both the AD-6

11

Above, VA-65 AD-6 refuels a VF-33 F11F Tiger in 1962. (NMNA) Below, VA-65 AD-6 BuNo 139678 from CVA-11 refuels an F4D-1 Skyray from VF-74 on 3 August 1960. (USN) Bottom, three VA-65 AD-6s BuNos 139730 (AF/403), 139734 (AF/406), and 139764 (AF/405) escort CAG-bird BuNo 137567 (AF/400) on 13 August 1960 while assigned to the Intrepid. CAG's rudder was red-blue-orange-green-purple and black. (USN)

and AD-4/4B operated on CVA-41 until it returned to Norfolk on 4 August 1954.

Nine more deployments were made with their AD-6/A-1Hs on three different ships before transitioning to the A-6A Intruder in November 1964. The first of these was to the Med aboard the USS Lake Champlain (CVA-39) from 9 October 1955 to 30 April 1956.

Four deployments aboard the USS Intrepid (CVA-11) followed with the first to the North Atlantic and the

other three to the Mediterranean. These were from: 3 September to 21 October 1957, 12 February to 30 August 1959, 4 August 1960 to 17 February 1961, and 3 August 1961 to 1 March 1962. Prior to the 1957 cruise, VA-25's tail code was changed to "AF" on 1 July 1957 and during the 1959 cruise the squadron was redesignated VA-65 on 1 July 1959. During this time, a twenty-day emergency deployment occurred in response to the assassination of GEN Trujillo in the Dominican Republic from 1 to 20 June 1961.

Above, VA-65 AD-5 BuNo 132463 had an orange rudder and engine cowl ring and da-glo orange vertical and horizontal tail. (Barry Miller) Below, four VA-65 A-1Hs BuNos 139744 (AE/411), 139714 (AE/410), 137538 (AE/407) and 139690 (AE/401) in flight off the USS Enterprise (CVAN-65) in June 1963. (Hal Andrews collection)

A ship and tail code change occurred in late 1962 when VA-65 was assigned to the nuclear powered Enterprise. CVG-6's new tail code

At left, VA-65 A-1H BuNo 135711 taxis on Enterprise in 1963 as part of CVG-6. Rudder was orange with white stars and fuselage lightning bolt was yellow. (Paul Minert collection) Below, VA-65 AD-6s BuNos 139758 (AE/403) and 139707 (AE/400) in 1964. (USN) Bottom, VA-65 A-1H BuNo 135305 traps aboard CVAN-65 in 1964. (USN)

became "AE". Four A-1H deployments were made aboard CVN-65. They were from: 3 August to 11 October 1962, 19 October to 6 December 1962, 6 February to 4 September 1963, and from 8 February to 3 October 1964. The first cruise was the ship's maiden voyage to the Med during which she participated in NATO exercise Riptide III. The second cruise was the emergency deployment due to the Cuban Missile Crisis where they participated in the blockade of Cuba. VA-65 returned to the Med for its third "Big E" cruise. Ports-of-call were: Cannes, Athens, Palermo, Sicily, Naples, Taranto, Genoa, Corfu, Rhodes, Beirut, and Barcelona. The last deployment was "Operation Sea Orbit", the 1964 around the world cruise of the all nuclear Task Force One which included the USS Bainbridge (DLGN-25) and the USS Long Beach (CGN-9). Ports-of-call were: Rabat, Dakar, Freetown, Monrovia, Abidjan, Nairobi, Karachi, Freemantle, Melbourne, Sydney, Wellington, Buenos Aires, Montevideo, Sao Paulo, Rio De Janeiro, and Recife.

While flying the Skyraider, the squadron received the Battle E in 1950, 1952, and 1960, the Atlantic Fleet Safety Award in 1956, and the CNO Safety Award in 1961. VA-65 was disestablished on 31 March 1993.

ATTACK SQUADRON THREE B, VA-3B
THE FIRST ATTACK SQUADRON FORTY - FOUR, VA-44 "GREEN KNIGHTS"

Bombing Squadron Seventy-Five (VB-75) was established at NAAS Chincoteague, VA, on 1 June 1945 with SBF-4Es. SB2C-4Es were received in August 1945 and SB2C-5s in March 1946. The squadron's duty station also changed in March to NAS Norfolk, VA, after VB-75 returned from deployment aboard the USS Franklin D. Roosevelt (CVB-42) on 19 March. A second FDR cruise

with Helldivers occurred from 8 August through 4 October 1946 and on 15 November VB-75 was redesignated Attack Squadron Three B (VA-3B).

LCDR Elmer Maul took command on 20 December 1946 and VA-3B received its first Douglas AD-1 Skyraider on 20 March 1947. By the end of June, the squadron was operating twenty-four AD-1s. One year later while under the command of LCDR Oscar I. Chenoweth, Jr., the squadron was still operating twenty-four AD-1s but had added two AD-1Qs to its complement. In July 1948, the squadron began operating from the FDR and on 1 September 1948 VA-3B was redesignated Attack Squadron Forty-Four (VA-44). VA-44 deployed to the Mediterranean aboard FDR with nineteen AD-1s and one AD-1Q from 13 September 1948 through 23 January 1949. After

returning to CONUS, LT K. F. Rowell took temporary command of the squadron until replaced by LCDR Robert N. Miller on 28 February 1949. VA-44 was transferred to NAS Jacksonville, FL, on 12 February 1949 where it acquired one Martin AM-1 and one AM-1Q. By July 1949, there were twenty AM-1s, one AM-1Q and one AD-1 aboard the USS Midway (CVB-41) for carrier qualifications. The Martin Maulers were withdrawn in October 1949 only to be replaced with eighteen AD-1s once again in November 1949. The squadron then made a Med deployment with their AD-1s aboard Midway from 6 January to 23 May 1950 as part of CVG-4. After the squadron's return to Jacksonville it was disestablished on 8 June 1950.

VA-3B AD-1 in flight in 1947. Note wing codes. (NMNA)

VA-3B/VA-44

Above, VA-3B AD-1 BuNo 09162 at the Cleveland National Air Races in September 1947. (Balough) At left, a blue tail stripe and propeller hub were added in late 1947 to VA-3B's AD-1s seen here with an APS-4 radar pod under the left wing. (Ginter collection) Below, four VA-3B AD-1s prepare to dive with open "barn door" belly dive brakes In 1947. Tail stripe was blue. (USN)

FLEET COMPOSITE SQUADRON THREE, VC-3

VC-3 was established on 2 May 1949, from Det One, Fleet All-Weather Training Unit, Pacific (FAW-TUPAC). The squadron transferred from NAS North Island to NAS Moffett Field, CA, in October 1949. At Moffett, the unit initially flew F6F-5Ns, TBM-3E/3Ns, two AD-2Qs, and an SNB-5. VC-3 was tasked with the mission of providing attack (including special weapons) and night fighter detachments to the Pacific Fleet. To fulfill this mission, the unit received more Skyraiders, F4U-5Ns, F8F-1Ns, F3D-2s, F2H-2Bs and F2H-3s.

By 31 October 1949, VC-3 had two AD-3Ns and two AD-2Qs. This grew to one AD-3, three AD-3Ns, three AD-3Qs, and two AD-2Qs by 31 January 1950. The first AD-4 arrived in February 1950 with two AD-4Ns and two AD-4Qs arriving in May. Also

in May 1950, the Attack Division (Skyraiders) were transferred back to San Diego where they were absorbed by the newly-established VC-35. VC-3's three Skyraider Dets remained in force until they were released from Korea on 7 April 1951.

During the Korean War, VC-3 sent seventeen detachments of F4U-5N nightfighters to the war zone. Three of these detachments also included Skyraiders. AD-3Ns were deployed on the USS Valley Forge (CV-45) from 1 May to 1 December 1950 (CAG-5) and again from 6 December 1950 to 7 April 1951

Above, the three Skyraider Dets to Korea were in concert with VC-3's F4U-5Ns. (USN via Peter Mersky) Below, VC-3 Team Charlie AD-3N BuNo 122900 with engine running prepares to taxi from the number two deck edge elevator to the foredeck for take-off on the USS Valley Forge (CV-45) on 5 February 1951 for a combat mission over Korea. (USN)

(CAG-2). The third Det with AD-4s was aboard the USS Philippine Sea (CV-47) as part of CVG-11 from 24 July 1950 to 7 April 1951.

Aircraft Development Squadron Three (VX-3) was established on 20 November 1948 at NAS Atlantic City, NJ, primarily from the equipment and personnel of CVLG-1 (VF-1L and VA-1L). CDR W.H. Keighley was the first commanding officer and had fifty-one aircraft at his disposal by 5 January 1949. These were F6F-5Ns, F8F-1s, F8F-2Ns, TBM-3Es, F4U-4s, F4U-5s, FH-1s, SNJ-4/5s, three XBT2D-1s and three AD-2s.

The squadron made refresher landings aboard the USS Saipan (CVL-48) from 30 November to 2 December 1948. On 9 December 1948, intensified instrument training began and the squadron was tasked to test projects with new ordnance equipment, pilot's flight clothing and cold weather operation of power plants. These projects were conducted aboard the USS Palau (CVE-122) on a cold weather cruise from 8 January to 28 January 1949. Three tests directly involved the Douglas

Skyraiders. The first tested the cold weather operation of the Douglas Bomb Ejector as installed in the ADs and resulted in it being released for fleet usage. The second test involved the 20mm gun heater as installed on AD aircraft, which was found unsatisfactory. And the third was cold weather starting and operation of the R-3350 engine. These tests were inconclusive as more adverse conditions were needed.

From 21 August to 1 September 1950, five AD-4Q aircraft were used to conduct the tactical evaluation of the bomb director Mk. 3 Mod 3 for fleet usage. The tests were conducted at NAOTS Chincoteague, VA. In November 1950, F4U-5s and AD-4Qs conducted 5.0 inch HVAR plaster-nosed radar ranging firing tests against drones shooting two down.

On 2 February 1951, LT J.E. Whillans bailed out of AD-4Q BuNo 124060 due to an engine fire. He landed about 3/4 of a mile off shore, but drowned due to exposure before being retrieved by a VP-661 PBM-5

Above, VX-3 AD-1 BuNo 09334 in flight. (USN) Bottom, VX-3 AD-4Q BuNo 124041 at NAS Glenview, IL, in July 1950. (Swisher collection)

out of Norfolk.

From 1 January to 30 June 1951, the AD was involved in developing tactics for maximum effectiveness of carrier aircraft against ground targets. ADs dropped thirty napalm bombs which resulted in tail fins being added to the tanks for increased stability during the drop. High altitude bombing tests were satisfactory from conventional dives from 20,000 ft, releasing at 14,000 as were tactics against railways using 250 lb land mines. The AD, F2H, F9F, and F4U aircraft were used to fire 429 HVAR and 53 SCAR rockets at airborne targets during this period, too.

From 25 August to 15 September 1951, VX-3 deployed aboard the USS Midway (CVB-41) with eight F4U-5s, ten F9F-2s, fourteen AD-1/2/4/4Qs,

and ten F2H-2s. During the cruise, VX-3 was tasked with developing tactics for maximum effectiveness of carrier aircraft against ground objectives. The ADs carried twelve 250 lb bombs, or six 500 lb bombs plus external drop tanks. During the deployment a couple of ADs were also used as target tugs.

VX-3 deployed eleven F2H-2/2P and twelve AD type aircraft aboard the USS Wright (CVL-49) from 6 to 12 January 1952 for carrier qualifications and project evaluations. Then the entire squadron deployed from 17 May to 6 June 1952 aboard the USS Leyte (CV-32) along with VF-171, VA-35, VC-12, and VS-27.

F2Hs and ADs deployed again from 11 to 15 August 1952, this time aboard the USS Saipan (CVL-48) for carrier qualifications and more project evaluations. They went to sea again

from 20 to 24 October 1952 for day and night qualifications in ADs, F2Hs, and F9Fs along with VX-5 aboard the USS Wasp (CVA-18). The final deployment of 1952 was from 24 October to 1 November aboard the USS Coral Sea (CVA-43) along with VX-5, VC-4 and VC-33.

And so it went test after test through December 1958 with VX-3 flying every version of the Skyraider.

Above, VX-3 AD-4N BuNo 127014 at Mines Field (LAX), CA, in 1953. (William T. Larkins collection) Below, two views of VX-3 AD-6 BuNo 135405 refueling a F3H-1 on 22 August 1958. (USN and USN via Jim Sullivan)

In 1957-58 the testing and development work weighed heavily on AD-6/7's tanker aircraft.

ATTACK SQUADRON FOUR B, VA-4B
THE FIRST ATTACK SQUADRON FORTY - FIVE, VA-45 "BLACK KNIGHTS"

The first VA-45 was originally established as Torpedo Squadron Seventy-Five (VT-75) on 1 June 1945 at NAAS Chincoteague, VA. VT-75 flew a mixed bag of SBF-4E, SBW-4E, and SB2C-4E Helldivers for a short training cruise aboard the USS Franklin D. Roosevelt (CVB-42) from 8 January to 19 March 1946. On return from the Caribbean, the squadron transferred to NAS Norfolk, VA, where they received SBC-5s prior to deploying to the Med aboard FDR from 8 August to 4 October 1946. After the cruise, VT-75 was redesignated Attack Squadron Four B (VA-4B) on 15 November 1946.

LCDR Chester L. Dillard took acting command on 19 February 1947 and in April began receiving AD-1 Skyraiders. By 2 May there were

seven AD-1s, eleven SBC-5s and four TBM-3Ws on-hand. LCDR Lucien G. Powell, Jr. relieved LCDR Dillard on 26 May and the squadron had a full complement of twenty AD-1s by the end of June. LCDR Powell was relieved by LCDR Leroy V. Swanson on 15 June 1948. The squadron was redesignated Attack Squadron Forty-Five (VA-45) on 1 September 1948 prior to its first Skyraider deployment to the Med aboard the FDR from 13 September 1948 through 23 January 1949 as part of CVG-4.

After returning to CONUS, VA-45 transferred to NAS Jacksonville, FL, on 14 February 1949 and in a interesting turn of fate transitions to the Martin AM-1 Mauler. In March, VA-45 had four AM-1s, nine AD-1s and one

Above, VA-4B AD-1 BuNo 09315 in 1947-48. Tail stripe was orange. (Jim Sullivan collection) Below, VA-45 AD-1 BuNo 09192 taxis at Wilmington, NC, in 1949. Note squadron insignia on the cowl, orange tail stripe and pilot's name, ENS Bill Newell, below the windscreen. (Jim Sullivan collection)

SNJ-5 on strength. Eventually, the squadron had a total of nineteen AM-1s in operation until it began re-equipping with AD-1s again in October 1949. The last AM-1 was transferred out in December leaving the squadron with eighteen AD-1s. After work-ups, VA-45 deployed aboard the USS Midway (CVB-41) from 6 January to 23 May 1950 to the Mediterranean. Upon return, LCDR Frederick C. Kidd assumed command of VA-45 prior to its disestablishment on 8 June 1950.

FLEET COMPOSITE SQUADRON FOUR, VC-4 "NIGHTCAPPERS"
ALL-WEATHER FIGHTER SQUADRON FOUR, VF(AW)-4

VC-4 was established at NAS Atlantic City, NJ, on 28 September 1948 as the AirLant night attack (including special weapons) and night fighter squadron. Their mission was to provide Dets to the Atlantic Fleet carriers. As such they equipped with F4U-5N/NLs, F2H-2N/-3/-4 and F3D-2 nightfighters. Special weapons day fighters in the form of F2H-2Bs, F9F-5Bs and F6F-6s were also acquired as was a mixed bag of Skyraiders.

The squadron's first Skyraiders were three AD-1Qs received in November 1948. By 31 January 1949, there were two AD-2Qs and nineteen AD-1Qs on-hand and in February the first Skyraider Det was sent aboard the USS Leyte (CV-32). In May, eight AD-1Qs deployed on the Coral Sea (CVB-43) and four went aboard Midway (CVB-41). Another four deployed in June aboard Kearsarge (CV-33) and eight aboard FDR (CVB-42) in October 1949.

In August 1950, all of VC-4's attack assets were absorbed in the newly established VC-33 except the four AD-3Ns of Det 5 aboard Midway which returned in November. In December 1955, Skyraiders returned to VC-4 in the form of twenty-two AD-5/5Ns which replaced all the squadron's jet aircraft. With the AD-5/5Ns came a new ASW mission in which Dets would provide protection to the ever-expanding number of Essex Class CVSs. The following Dets were deployed:

Jan 56	Det 38	CVS-40	Four	AD-5/5N
Feb 56	Det 52	CVS-45	Four	AD-5/5N
Mar 56	Det 50	CVS-36	Eight	AD-5/5N
Jan 57	Det 51	CVS-32	Three	AD-5/5N
Sep 57	Det 38	CVS-40	Four	AD-5/5N
Sep 57	Det 45	CVA-9	Four	AD-5/5N
Sep 57	Det 51	CVS-32	Four	AD-5/5N
Jan 58	Det 52	CVS-45	Four	AD-5/5N

On 2 July 1956, VC-4 was redesignated VF(AW)-4 and F2H-3/4 were acquired beginning in 1957 which operated in separate Dets from those of the Skyraiders. On 1 July 1957, VC-4's tail code changed to "GC".

Above, VC-4 AD-2Q Skyraider NA/94. (via Tailhook) Below, VC-4 AD-1Q Det 4 NA/7 takes the barrier aboard the USS Kearsarge (CV-33) in June 1949. (Dave Benton via Larry Webster) Bottom, VF(AW)-4 AD-5/5N landing aboard the USS Leyte (CVS-32) in February 1958 with new "GC" tail code. (USN)

The last AD-5 was retired in August 1958 and the squadron was disestablished in July 1959.

UTILITY SQUADRON FOUR, VU-4 "DRAGON FLYERS"

VU-4 was originally established as VJ-4 on 15 November 1940 at NAS Norfolk, VA. Subsequent home bases were Guantanamo, San Juan, Squantum, Chincoteague and Oceana. The squadron operated over thirty types of aircraft from the PBY-1 Catalina to the supersonic F-8 Crusader.

The mission of VJ-4 during WWII was to fly anti-submarine patrol and conduct rescue operations. After the war, the squadron was redesignated Utility Squadron Four (VU-4). Its mission was changed to fleet support in air-to-air and surface-to-air weapons training as well as training fleet air intercept controllers and radar and ECM operators.

In 1952, the squadron added F6F-5K and F9F-5K drones to its inventory for use as realistic targets for the fleet. the squadron maintained a couple of Skyraiders as target tow and spotter aircraft through 1958. In June 1958, P2V-5 and UB-26J aircraft were added to launch KDA drones. VU-4 was redesignated VC-4 on 1 July 1965.

Below, VU-4 AD-4N BuNo 124146 with "UD" tail code is seen here at Chincoteague in target tow colors in 1956. (Roger Besecker) Bottom, retired VU-4 AD-4NA BuNo 126940 at NAF Litchfield Park, AZ, on 28 February 1958. (NMNA)

CARRIER QUALIFICATION UNIT FOUR, CQTU-4

Carrier Qualification Unit Four (CQTU-4) operated at NAAS Correy Field, Pensacola, FL, and nearby NAAS Barin Field. During the late 1940s and early 1950s, the unit operated TBMs, F8Fs, F6Fs, and AD Skyraiders. CQTU-4 received its first two AD-1s in November 1949. It would utilize up to twenty AD-1s and AD-3s until its retirement in January 1953 at Barin Field.

Above, CQTU-4 AD-1s with "CB" tail code during Carrier Qualifications. (USN) Below, CQTU-4 AD-1 (KA/62) taxis at Barin Field, FL, in 1949. (NMNA) Bottom, CQTU-4 AD-1 (CB/62) at NAAS Corry Field, Pensacola, FL, in 1950. (NMNA)

ATTACK SQUADRON FIVE B, VA-5B "BLACK LANCERS"
ATTACK SQUADRON SIXTY - FOUR, VA-64 "BLACK LANCERS"
COMPOSITE SQUADRON TWENTY - FOUR, VC-24

VA-5B   VA-64

Above, CDR Paul Buie makes the first take-off from CVB-43 in a VA-5B AD-1 C/311 in January 1948. (USN) Below, VA-64 AD-1Q (C/400) at the Cleveland National Air Races in September 1948 While assigned to CVBG-6 aboard the Coral Sea. It was armed with rockets, bombs and an APS-4 radar unit. (Warren Bodie)

Bombing Squadron Seventeen (VB-17) was established on 1 January 1943. After WWII, VB-17 was redesignated Attack Squadron Five B (VA-5B) on 15 November 1946 while flying Curtiss Helldivers.

The squadron received its first AD-1 Skyraider in October 1947. By 15 December 1947, VA-5B had a full complement of twenty-four AD-1s while still operating eight SB2C-5s and one SNJ-4. VA-5B was redesignated Attack Squadron Sixty-Four (VA-64) on 27 July 1948.

VA-5B participated in the shake-down cruise of the USS Coral Sea (CVB-43) as part of CVBG-6 from 19 January through 5 April 1948 during which CDR Paul Buie made the ship's first take-off and landing in a VA-5B Skyraider.

VA-5B/VA-64 then deployed aboard CVB-43 to the Caribbean in May and June 1948 before the Coral Sea left on a Midshipman cruise with CVEG-2 from 7 June to 6 August 1948. Carrier qualifications were later conducted in September 1948 aboard

CVB-43, in November aboard CVB-41, and in February 1949 aboard CVB-41 again. In March 1949, CVG-6 was transferred to NAS Oceana, VA, from NAS Norfolk, VA, along with VA-64 and sister Skyraider squadron VA-65.

On 8 April 1949, VA-64 was redesignated Composite Squadron Twenty-Four (VC-24) and had two AD-1s on-hand on 31 April. Only TBM-3Es were on-hand in June 1949. VC-24 was in turn redesignated

Anti-Submarine Squadron Twenty-Four (VS-24) on 20 April 1950. VS-24 was disestablished on 1 January 1956.

Above, VA-64 AD-1 BuNo 09283 C/417 being catapulted from the USS Coral Sea CVB-43 on 14 September 1948. Fin tip was yellow. (National Archives) Below, VA-64 AD-1 C/409 unfolds its wings as it taxis forward to the catapults on CVB-43 on 14 September 1948 (NMNA).

AIR DEVELOPMENT SQUADRON FIVE
AIR TEST AND EVALUATION SQUADRON FIVE, VX-5 "VAMPIRES"

VX-5 was established on 18 June 1951 at NAS Moffett Field, CA. The original mission was to develop and evaluate aircraft tactics and procedures for the delivery of airborne special weapons. The squadron moved to NAS China Lake in July 1956 where the mission expanded to the operational test and evaluation of all new air-to-surface weapon systems. To prosecute its mission VX-5 operated all fleet tactical aircraft.

Above, VX-5 formation of F3D, AD-4, F9F, F2H-2 and F2H-4. (National Archives) Below, three VX-5 Skyraiders in flight off San Francisco on 13 March 1953 with AD-4B BuNo 127854 in the foreground. (National Archives)

VX-5 "VAMPIRES"

Above, VX-5 AD-5N BuNo 132488 with two more ADs at left and the nose of a F3D-2 and a F7U-3 at right. (Ginter collection) At right, Operational Test and Evaluation Force A-1E BuNo 133922 had passenger seats in the aft compartment and a green tail stripe. (Ginter collection) Below, VX-5 AD-7 BuNo 142011 during carrier qualifications on the USS Bennington (CVA-20) in May 1959. Cowl and tail stripes were green outlined by thin white stripes. (USN)

Above, VX-5 AD-7 with shape during a photo op with squadron A4D-2N BuNo 148565 (via Tailhook) Below, VX-5 AD-7 BuNo 142015 on a sandblower low level navigation hop with a nuclear shape. (NMNA)

UTILITY SQUADRON FIVE, VU-5 "WORKHORSE OF THE FLEET"

Fleet Utility Squadron Five (VU-5) was established on 16 August 1950 in order to provide utility services to the fleet in and around the islands of Japan. As the squadron and its services expanded, UTRON-5 was required to send detachments to carriers and bases as far from its home base of NAS Atsugi as NAS Cubi Point, Philippines. VU-5's primary mission was to provide targets for both the aerial and surface components of the fleet. Other missions consisted of photo services, adversary missions, airborne control of surface launched missiles, and carrier on-board delivery (COD) services.

Only the first AD-5 received was operated by Det B Naha in October 1959. All other Skyraiders in service with VU-5 were AD-5/A-1Es and were assigned to VU-5 Det A Cubi from November 1959 through June 1964. The average number of AD-5/A-1Es on-hand was three up until January 1964. One A-1E only was on-hand from January through June 1964.

Above, two VU-5 P2V Neptunes, one JD Invader and two AD-5s in flight in late 1961. (USN via Angelo Romano) Below, VU-5 aircraft numbers 21 through 25 on the Det A A1-E line at Cubi Point, PI, with BuNo 133898 in the foreground on 25 April 1963. Aircraft were painted in target tow colors of engine grey with yellow wings and horizontal tail surfaces and da-glo red vertical tail and wing stripes. (NMNA)

ATTACK SQUADRON SIX B, VA-6B
THE FIRST ATTACK SQUADRON SIXTY-FIVE, VA-65 "FIST OF THE FLEET"
SECOND ATTACK SQUADRON TWENTY-FIVE, VA-25 "FIST OF THE FLEET"

VA-25 was originally established as Torpedo Squadron Seventeen (VT-17) on 1 January 1943 at NAS Norfolk, VA, with TBF-1 Avengers. During WWII, VT-17's finest moment was being one of the torpedo squadrons responsible for sinking the mighty 64,000 ton Yamato. The unit was credited with four torpedo hits.

In March 1946, the Avengers were replaced with SB2C Helldivers at NAS Brunswick, MN. The squadron relocated to Norfolk on 15 August 1946 where they were redesignated Attack Squadron Six B on 15 November 1946. While assigned to CVBG-6, VA-6B received its first AD-1 on 23 September 1947 and by the end of December twenty-three were on hand. Some of the squadron's Helldivers lingered on into February 1948.

VA-6B participated in the shakedown cruise of the USS Coral Sea (CVB-43) from 19 January through 15 April 1948. The squadron was redesignated Attack Squadron Sixty-Five (VA-65) on 27 July 1948 and deployed to the Mediterranean aboard the USS Midway (CVB-41) from 4 January to 5 March 1949 with sister Skyraider squadron VA-64. The two AD units shared the deck with two Marine F4U-4 squadrons, VMF-

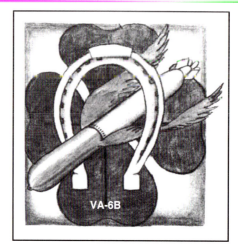

VA-6B

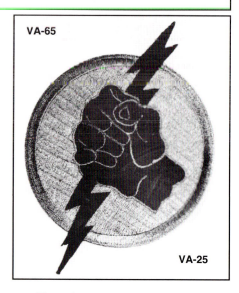

VA-65

VA-25

225 and VMF-461, and Navy F4U-4 squadron VF-172. After returning, the squadron was relocated to NAS Oceana and a short four week deployment aboard the USS Franklin D. Roosevelt (CVB-42) to conduct cold weather exercises in the Davis Straits was conducted from 27 October to 23 November 1949. The following month they began receiving AD-4s and had eighteen on-hand by the end of the month. The squadron adopted the nickname "Fist of the Fleet" on 25 June 1950.

The squadron was awarded the AIRLANT Battle Efficiency "E" for single seat attack squadrons while aboard Midway in 1948 and again in 1950.

Transferred to the Pacific due to the Korean War, the squadron deployed to Korea while under command of CDR R.W. Phillips in the fall of 1950 as part of Air Group Two aboard the USS Boxer (CV-21). During the cruise ENS R.R. Sander's AD-4 was hit by small arms fire, the engine seized, and an emergency crash landing was made on a road near Seoul after cutting down a tree with the Skyraider's wing. The

Below, VA-6B AD-1 at Guantanamo Bay, Cuba, in February 1948 during the shakedown cruise of the Coral Sea. (Stan Kalas, Tom Doll collection)

Three views of VA-6B flight operations aboard the USS Coral Sea (CVB-43) on 20 May 1948. Note horseshoe insignia on the nose of all the squadron's AD-1s. (National Archives)

squadron's first combat tour was a short one, lasting from 24 August to 11 November 1950. Three weeks later they boarded the USS Valley Forge (CV-45) in San Diego on 6 December 1950 while still under command of CDR R.W. Phillips. The "Happy Valley" joined TF-77 on 22 December and commenced air strikes the next day. VA-65 operated on CV-45 until 28 March 1951 and

Above, VA-65 AD-2 BuNo 122269 landing at NALF Santa Rosa, CA, in August 1951. Fin tip was green. (William T. Larkins) Below, VA-65 AD-4 123851 deck launches from CV-47 for an attack over Korea in 1951. (Tailhook) Bottom, VA-65 AD-3 taking off from the USS Valley Forge (CV-45) in April 1951 loaded with three 2,000 lb bombs. Note the two gunnery pennants (green with red dot) on the side of the fuselage below the canopy. (National Archives)

then was crossdecked to the USS Philippine Sea (CV-47). While flying from Boxer, Valley Forge, and Philippine Sea in Korea, VA-65 flew 1,718 missions, 5,538 combat hours, lost five planes and two pilots. They operated over Korea until 9 June 1951, before returning to CONUS and being re-assigned to NALF Santa Rosa, CA. After leave, new personnel, and lots of training, VA-65 conducted carrier qualifications aboard

Above, VA-65 AD-4 BuNo 128925 chewed up a VF-24 F4U-4 on the deck of CV-31 on 22 January 1952. (National Archives via Jim Sullivan) Below, VA-65 AD-2 BuNo 122226 being lowered onto the Philippine Sea (CV-47) at Yokosuka, Japan, in 1951. (NMNA)

the USS Bon Homme Richard (CV-31) in late January 1952. On 8 February 1952, the squadron sailed aboard the Boxer for its last Korean War combat deployment. The squadron participated in a coordinat-ed four carrier strike on North Korean hydro-electric power plants with VA-65 striking the Suiho power plant on the Yalu River. CDR Gordon Sherwood was in command through-out the cruise which ended on 26

Above, VA-65 AD-4 BuNo 128920 about to deck launch from CV-21 in 1952. (Harry Gann) Below, VA-65 Skyraider being re-spotted aft by muscle power on 18 June 1952 aboard the USS Boxer (CV-21). Aircraft was carrying twelve incendiary bombs. (National Archives)

September 1952. Upon return to CONUS, the unit was transferred to NAS Alameda, where it would re-group and train for its next WestPac deployment.

Nearly a year later, on 3 August 1953, VA-65 sailed westward aboard the USS Yorktown (CVA-10) with its AD-4B/AD-4NAs. This deployment ended on 3 March 1954. Eight months later, VA-65 was aboard the USS Essex (CVA-9) where the squadron participated in the evacua-

Above, VA-65 AD-4NA flown by LTJG Charles Carter nosed over on landing aboard the USS Yorktown (CVA-10) on 20 January 1954. (USN) Below, VA-65 flight line at NAS Alameda in March 1954. Aircraft have green fin tips and green and white rudder control tabs. (USN)

tion of the Tachen Islands near Formosa and then monitored the area until returning to Alameda on 21 June 1955 while under the command of CDR John R. Bowen II.

It wasn't until 13 November 1956, some sixteen months later, that the squadron deployed again. Still assigned to CVG-2 they sailed aboard the USS Shangri-La (CVA-38) while equipped with new AD-6s. The biggest exercise during this deployment was Operation Beacon Hill, a simulated invasion of the Philippines. Upon returning to Alameda on 22 May 1957, CDR William Bates relieved CDR Bowen.

Still stationed at Alameda, training Dets were conducted at Fallon and China Lake for both conventional and special weapons delivery. Then on 16 May 1958, CarQuals for the upcoming Midway deployment were conducted aboard the USS Kearsarge (CVA-33). The Midway cruise was from 16 August 1958 to 12 March 1959 and was highlighted by the participation in the Formosa Crisis and the threatened occupation of the

At left, VA-65 pilots and planes aboard the USS Essex (CVA-9) during CVG-2's 1956-57 WestPac deployment. (USN) Below, VA-65 AD-6 BuNo 137618 was assigned to the USS Essex when seen at NAS Oakland on 18 September 1955. Cowl and tail trim was green and white. (William T. Larkins)

Quemoys and Matsu Island, the Pescadores and Taiwan by the Chinese Communists. After the cruise, VA-65 was redesignated Attack Squadron Twenty-Five (VA-25) on 1 July 1959.

A quick turn-around was made

and VA-25 deployed on Midway with AD-7s from 15 August 1959 to 25 March 1960. This was followed by another Midway cruise from 16 February to 20 October 1962.

6 April 1962 found VA-25 aboard Midway with four AD-7s and eight

Above, VA-65 CAG bird AD-5 BuNo 133929 at NRAB Minneapolis, MN, in 1956. (Jim Sullivan collection) Below, VA-65 AD-5 BuNo 132646 had green trimmed engine cowl and tail. (Balogh via Menard) Bottom, VA-65 AD-6 BuNo 137561 on board the USS Shangri-La (CVA-38) in 1957. Engine and tail trim were green. (USN)

Above, Midway-based VA-25 AD-6 BuNo 137627 landing at NAF Atsugi, Japan, on 2 August 1962. Fin tip was green. (Toyokazu Matsuzaki) Below, two Coral Sea-based VA-25 A-1Hs in 1966 with BuNo 135329 at left. Fin tips were black. (Paul Minert collection) Bottom, VA-25 Skyraiders BuNos 137495 NE/505 and 142037 NE/502 taxing out for a training mission at NAS Lemoore, CA, in 1965, were assigned to the USS Midway (CVA-41). (Tailhook)

AD-6s headed toward WestPac and Operation Counter Thrust in June. An amazing 1,031 flight hours were amassed that month. Midway and VA-25 returned to the states on 20 October 1962.

Almost thirteen months later, on 8 November 1963, the squadron deployed on Midway again. CVA-41 joined with the USS Kitty Hawk (CVA-63) for a head-to-head competition at sea, during which VA-25 defeated VA-115 in bombing and rocket firing. The squadron returned to NAS

Lemoore, CA, on 26 May 1964.

A Hawaiian "Pineapple Cruise" aboard Midway was conducted from 6 October to 9 November 1964 which included a simulated strike on California. Twice more VA-25 attacked California, in February and March 1965 during simulated invasions.

The Fists' next cruise was to Vietnam aboard Midway. They sailed on 6 March 1965 and arrived in the Gulf of Tonkin on 10 April after strike training off Luzon. VA-25 became the first Midway squadron to engage the enemy in battle when a flight of four led by LCDR H.E. Gray were conducting a ResCap mission in Laos on 12 April. From 13 April to 9 June, the squadron hit anything anywhere that moved or didn't move of value. In South Vietnam they attacked Viet Cong strongholds near Saigon and along the Mekong Delta, and covered the amphibious landings at Chu Lai. In the North they hit the Bai Duc Thon highway bridge, truck convoys, PT

boats, supply depots, and Vinh Harbor.

On 24 May, CDR Ralph Smith was relieved by CDR Harry Ettinger and seventeen days later the squadron struck the Co Dinh thermo-electric plant. Eight aircraft carrying 2,000 lb bombs and led by CDR Stoddard made two runs each while four A-4s provided flak suppression. LTJG J.S. Lynne was credited with bringing down the facility, but not without consequences. LTJG C.L. Doughtie, flying Canasta 578 (A-1H

Above, VA-25 AD-7 BuNo 142023 in flight. (Ginter collection) Bottom, VA-25 A-1H BuNo 137627 in flight while assigned to CVA-41. Fin tip was green. (Paul Minert Collection)

BuNo 137521), failed to return after his last pass.

On 20 June 1965, four VA-25 A-1Hs led by LCDR E.A. Greathouse were dispatched on a RESCAP mission into North Vietnam. At 12,000ft and 50-miles northwest of Thanh

Above, Lt Clint Johnson's MiG killer A-1H BuNo 139769. (Jim Sullivan collection) Below, VA-25 MiG killer, LT Clint Johnson, posing on 17 March 1967 next to BuNo 135294 at Lemoore. Johnson shared credit with Charles Hartman for the 20 July 1965 shootdown. (William Swisher) Bottom, bombed-up VA-25 A-1H BuNo 137517 on the deck of the USS Coral Sea (CVA-43) awaiting a mission over Vietnam in 1967. (Jim Sullivan collection)

Hoa, the flight was jumped by two MiG-17s. The A-1s dove for the deck, jettisoning their fuel tanks and ordnance, while turning inside the MiGs thus denying them any hits. Greathouse with LTJG J.S. Lynne on his wing broke off with one MiG chasing after them. The other two Skyraiders flown by LT Clint Johnson and LTJG Charles Hartman began an attack on the MiG which then turned to meet them head-on. Both pilots began scoring hits on the MiG, which rolled inverted and crashed. The second MiG bugged-out with the A-1s in pursuit. Each pilot was credited with a half a kill and received a Silver Star for the event. Johnson flew BuNo 139768 (NE/577) and Hartman flew BuNo 137523 (NE/573). Greathouse and Lynne were awarded DFCs for their efforts. During the cruise, VA-25 was awarded the 1964-65 NAVAIRPAC Battle "E".

On 24 June 1965, an eight plane strike into North Vietnam was led by CDR Ettinger. A-1H, BuNo 137523,

was lost that day when its engine quit on the return to the ship. LCDR R.L. Bacon was rescued by a UH-2 from Midway. On 27 June, the squadron finished its period on the line by dropping a mountain side across a supply road. The squadron returned to the line on 23 July and dropped four bridges that day. On the 24th, LT Abe Abrahamson and LTJG Nick Daramus got in a successful dual RESCAP mission. They first covered the rescue of a F-105 pilot in Laos where they expended some of their 2.75" FFARs and 20mm. Then, as they were pulling out they were vectored to a downed A-6 crew. The pilot and BN were down in the rice paddies about a half mile apart and were being surrounded by approaching Viet Cong. The A-1s had held off the enemy for two hours and expended most of their ammo and rockets when notified of the approaching Marine rescue helos. The helos made successful pickups in each case and whisked them away. Total mission time to save three airmen was 7.8

hours. More successful RESCAPS were conducted on 27 July and on 28 July.

On 7 August 1965, a strike was diverted to the Dan Hoi Citadel where LCDR H.E. Gray was shot down in A-1H, BuNo 135329, by 37mm AAA fire. He went down in the target area on his second run and was killed in action. Despite the loss, VA-25 hit Dien Bien Phu and Dong Hoi in the days that followed.

The squadron's XO, CDR Stoddard, was wounded on 15 September by AAA during a RESCAP mission. He diverted to Da Nang where he made a wheels-up landing in A-1J BuNo 142021. More successful RESCAPS were conducted on 20 August and 24 September.

The month of October was filled

Above, VA-25 A-1H BuNo 137593 enroute to a mission over Vietnam in 1966. (NMNA) Bottom, VA-25 A-1Hs BuNos 135275 and 137517 escorting an Albatross in 1967. (NMNA)

with armed strike missions. During one of these missions on 20 October, four squadron A-1s saved the Special Forces camp at Plei Me from extinc-

tion with their accurate close air support. On the 23rd, the road thirty miles south of Plei Me was hit by CDR Ettinger and two other VA-25 Skyraiders. The result was 102 VC killed and their equipment destroyed. Attack on a VC compound near Saigon followed which left some 22 buildings destroyed.

On the squadron's last operational day off Vietnam, the recovered CDR Stoddard delivered the 6,000,000th pound of ordnance dropped by VA-25 during the deployment. It was a very special weapon, a toilet! After this feat, the Midway turned for home on 4 November 1965.

For the squadron's second war cruise, they sailed on the USS Coral Sea (CVA-43) from 29 July 1966 to 23 February 1967. CDR Stoddard who was XO on the Midway cruise took command of the squadron for the deployment. On the second day of combat operations, 14 September, Stoddard's A-1H, BuNo 139756 (NE/580), was shot down by an SA-2 SAM missile during a strike against a storage facility near Vinh. That afternoon, the XO CDR James Burden took command of the squadron.

Because of the SAM threat, the squadron's main missions became that of armed reconnaissance and RESCAP. On 12 October, LT Deane Wood's A-1H, BuNo 135323 (NE/572), was hit by AAA on a recon mission along Route 15. During his attempt to reach the coast, the wing caught fire and he bailed out over enemy territory. For the next week several air and ground attempts were made to recover Woods, but they all failed and he was captured. He became a POW and was released on 4 March 1973.

During the squadron's second line period, beginning on 4 December 1966, the mission of aerial spotters for destroyers north of the DMZ was added. The Coral Sea departed for San Francisco on 11 February 1967. During the squadron's second war deployment they flew 1,184 combat missions from the USS Coral Sea.

At right, VA-25 A-1J 142077 with CVW-15's "NL" tail code taxis aboard Coral Sea in 1968. (Paul Minert collection) Bottom, VA-25 A-1H BuNo 135356 at NAS Alameda, CA, in 1968. (William Swisher)

The squadron's third war cruise with the A-1 would be the last deployment of the Skyraider with an attack squadron. It was conducted aboard the Coral Sea again, but this time as part of CVW-15. CDR Burden was still in command when CVA-43 deployed on 26 July 1967, but would be relieved by CDR Clifford E. Church on 1 October 1967.

On 19 August, LCDR Fred Gates was lost at sea in A-1H, BuNo 137575, due to engine troubles. He crashed about a quarter mile aft of the ship while attempting to land. Another aircraft was lost due to engine problems on 30 August. LTJG Larry Gardiner was making a rocket run in A-1H, BuNo 135390 (NL/412), on an enemy logistics boat off North Vietnam when his engine quit and he ditched at sea. An Air Force H-53 "Jolly Green Giant" made a successful recovery of Gardiner amidst shore battery fire that was being silenced by five other VA-25 Skyraiders.

Another dramatic rescue was

conducted during the squadron's second line period. Two EA-3B crewmen had bailed out of a their flaming aircraft and landed amidst a North Vietnamese fishing fleet. VA-25's LCDR Smith assumed the role of on-scene commander and then orchestrated six highly accurate rocket and strafing runs that persuaded the fishing fleet to depart the area. While the SAR helo "Clementine Two" was escorted to the scene by LCDR Bolt, the enemy shore batteries opened fire. Their involvement was short lived as two assisting A-4s attacked them. Both crewmen were successfully rescued.

On 4 December 1967, the destroyer USS Ozbourn (DD-846) was hit by a coastal battery north of

Cape Mui Ron, wounding several crewmen. VA-25's LTJG Thom and ENS Ramsey were ordered to assist and attacked the site with numerous rocket runs on the gun emplacement. On 26 December, LTJG Marcus was hit by automatic weapons fire while attacking ten enemy surface craft. His aircraft returned with a holed prop blade and a shattered windscreen, but the ten enemy ships were sunk. Then on 29 December the squadron provided RESCAP for the recovery of a VF-161 pilot and RIO. On 26 January 1968, the squadron aided in the evacuation under fire of a Laotian outpost by keeping the NVA troops at bay with repeated ordnance drops and 20mm runs. On 7 February, LCDR "Zip" Rausa and LTJG Gardiner were called to the aid of

eight Special Forces troops trapped at the Lang Vei Special Forces camp. The two Fist pilots strafed and bombed the northern half of the camp and were hit by small arms fire. An hour later, CDR Church, LT Thom, LTJGs Pellot and Hagen continued the close air support, and then the Air Force came in followed again by VA-25 in the late afternoon. This time it was A-1s flown by CDR Smith, LCDR Bolt, LT Jordan and ENS Hill. As the light was fading and all weapons had been expended, seven of the eight escaped, with the eighth wounded man being rescued later.

VA-25 returned to Lemoore on 6 April 1968 and retired the last attack squadron Skyraider, A-1H BuNo 135300, on 10 April. It was flown away by LTJG Tex Hill, who had made the last combat A-1 trap on 26 February 1968. It can be seen today

Above, VA-25 A-1J BuNo 142076 at NAS Miramar while assigned to the USS Coral Sea (CVA-43) in 1968. (William Swisher) Below, last fleet attack squadron A-1H BuNo 135300 Skyraider runs-up for its retirement flight at NAS Lemoore, CA, piloted by LT Tex Hill. (Ginter collection)

on display at the National Museum of Naval Aviation.

COMPOSITE SQUADRON ELEVEN, VC-11
AIRBORNE EARLY WARNING SQUADRON ELEVEN, VAW-11 "EARLY ELEVEN"

Carrier Airborne Early Warning Squadron One (VAW-1) was established on 6 July 1948 with twelve TBM-3Ws, ten TBM-3Es, one SNJ and one SNB. With these assets, the AEW Team/Detachment system of three to four aircraft per carrier was born. VAW-1 was redesignated Composite Squadron Eleven (VC-11) on 1 September 1948. While stationed at NAS San Diego, CA, VC-11 received its first Skyraiders, four AD-3s and four AD-3Ws in March 1949.

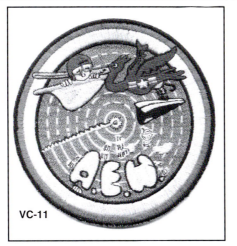

VC-11

When the Korean War was declared, VC-11 Team Five was aboard the USS Valley Forge (CV-45) with three AD-3Ws and thus became the first VC detachment to see combat. During the early stages of the war, the squadron's AD-3Ws and later AD-4Ws were utilized mostly in the ASW role, not the AEW role. VC-11 also received a small number of AD-3Qs and AD-4Qs.

Korean War VC-11 deployments:

CARRIER	DATES	OIC
CV-45 Det	01 May 50-01 Dec 50	LCDR Shelton
CV-47 Det	24 Jul 50-07 Apr 51	
CV-21 Det A	24 Aug 50-11 Nov 50	LT Dunning
CV-37 Det	09 Nov 50-29 May 51	LT Lynn
CV-45 Det	06 Dec 50-07 Apr 51	
CV-21 Det F	02 Mar 51-24 Oct 51	LCDR Haley
CV-47 Det	28 Mar 51-09 Jun 51	
CV-31 Det G	10 May 51-17 Dec 51	LT Kirk Jr.
CV-9 Det B	26 Jun 51-25 Mar 52	LCDR Miller
CV-36 Det D	08 Sep 51-02 May 52	LCDR Conyers
CV-45 Det H	15 Oct 51-03 Jul 52	LT Hyde
CV-47 Det C	31 Dec 51-8 Aug 52	LCDR Botten
CV-21 Det A	08 Feb 52-26 Sep 52	LT Waddell
CV-37 Det E	21 Mar 52-03 Nov 53	
CV-9 Det I	16 Jun 52-06 Feb 53	LCDR Knight
CV-33 Det F	11 Aug 52-17 Mar 53	LT Riggan
CV-34 Det G	15 Sep 52-18 May 53	LT Gernert
CV-45 Det B	20 Nov 52-25 Jun 53	LCDR Meredith
CV-47 Det M	15 Dec 52-14 Aug 53	LT Wortman
CV-37 Det D	24 Jan 53-21 Sep 53	LT Pierce
CV-21 Det H	30 Mar 53-28 Nov 53	LCDR Norton

Note: the CV designation changed to CVA on 1 Oct. 1952 but is not listed as such above due to lack of space.

On 14 September 1951, an AD-4W flown by LT B.E. O'Brien of VC-11 Det B

At top right, VC-11 AD-4W at Barbers Pt., TH, on 16 November 1951. (USN) Above right, VC-11 AD-3W being pushed back out of the barrier in 1951 aboard CV-31. Aircraft has a broken right wing and bent prop. (USN) At right, VC-11 AD-4W ND/36 "Faith" taxiing over the barrier on CVA-34 in 1955. (USN)

Above, two VC-11 AD-3Ws ND/39 and ND/84 from Boxer were diverted to an emergency field after being hit by AA on 13 June 1951. (National Archives via Jim Sullivan) At left, VC-11 AD-4Q BuNo 124124 impacts the Island on CVA-10 in 1954. (NMNA) Bottom, VC-11 AD-4W BuNo 124764 in flight from the USS Essex (CV-9) in 1952. (Ginter collection)

aboard Essex was lost. On take-off, the catapult bridle snagged on the tail hook and wrapped around the horizontal stabilizer creating continual buffeting and limited controlability. This precluded a carrier recovery and he was forced to ditch. O'Brien put her down near the USS Cony (DDE-508) at night. The pilot, who was weak from fighting the buffeting, was helped out of the cockpit by his two crewmembers and after some effort and time in the heavy swells they were picked up by the destroyer.

In April 1954, the first four AD-5Ws were received. However, this much more capable version did not fully replace the AD-3W until April 1955 and the AD-4Ws until February 1956. The last AD-4Q left the squadron in November 1956, leaving only the AD-5Ws. The unit operated without any "Q" birds until December 1958 when one AD-5Q was acquired.

On 2 July 1956, VC-11 was redesignated Airborne Early Warning Squadron Eleven (VAW-11). It then became one of the largest squadrons in the Navy. In January 1959, the unit had forty-nine AD-5Ws, twenty-three

Above, VC-11 AD-4W on 28 September 1956. Note unusual location of the VC-11 designation. (USN) Below, VAW-11 AD-4Q BuNo 124038 in 1956. (William Swisher)

AD-5Qs, twelve F2H-4s and one SNB on hand. The "Big Banjos" provided protective Sidewinder armed fighter Dets for the ASW carriers in 1958-59. There were fourteen ASW carriers in commission in July 1959 when VAW-11's tail code "ND" was replaced with

Above, VC-11 Det J cruise patch art-work for deployment aboard the USS Shangri-La (CVA-38) in 1956. Above left, VC-11 AD-5 BuNo 132641 at the Dayton Airport on 3 September 1954. (Ginter collection) At left, VC-11 Det J AD-5W BuNo 135173 aboard CVA-38 in 1956. (USN) Below, VAW-11 AD-5Ws in flight over the USS Yorktown (CVA-10) on 17 June 1957. (National Archives)

VC-11:

Jan 54	AD-3 (1), AD-4W (25)
Feb 54	AD-3W (2), AD-4W (25)
Mar 54	AD-3W (2), AD-4W (22)
Apr 54	AD-3W (2), AD-4W (21), AD-5 (4)
May 54	AD-3W (3), AD-4W (15), AD-5 (6)
Jun 54	AD-3W (3), AD-4W (12), AD-5 (6)
Jul 54	AD-3W (2), AD-4W (12), AD-5 (5)
Aug 54	AD-3W (3), AD-4W (15), AD-5 (5)
Sep 54	AD-3W (3), AD-4W (13), AD-5 (5)
Oct 54	AD-3W (3), AD-4W (13), AD-5 (5)
Nov 54	AD-3W (3), AD-4W (13), AD-5 (5)
Dec 54	AD-3W (1), AD-4W (13), AD-5 (5), AD-5W (7)
Jan 55	AD-3W (1), AD-4W (10), AD-5 (5), AD-5W (9)
Feb 55	AD-3W (1), AD-4W (10), AD-5 (5), AD-5W (9)
Mar 55	AD-3W (1), AD-4W (7), AD-5 (5), AD-5W (9), AD-4Q (1)
Apr 55	AD-4W (8), AD-5 (4), AD-5W (13), AD-4Q (2)
May 55	AD-4W (11), AD-5 (4), AD-5W (22), AD-4Q (4)
Jun 55	AD-4W (7), AD-5 (2), AD-5W (29), AD-4Q (4)
Jul 55	AD-4W (4), AD-5W (39), AD-4Q (3)
Aug 55	AD-4W (2), AD-5W (44), AD-4Q (2)
Sep 55	AD-4W (3), AD-5W (48), AD-4Q (3)
Oct 55	AD-4W (2), AD-5W (37), AD-4Q (1)
Nov 55	AD-4W (5), AD-5W (37), AD-4Q (1)
Dec 55	AD-4W (4), AD-5W (37), AD-4Q (1)
Jan 56	AD-5W (31)
Feb 56	AD-5W (36), AD-4Q (1)
Mar 56	AD-5W (37), AD-4Q (1)
Apr 56	AD-5W (34), AD-4Q (1)
May 56	AD-5W (33), AD-4Q (2)
Jun 56	AD-5W (35), AD-4Q (2)

VAW-11:

Jul 56	AD-5W (33), AD-4Q (1)
Aug 56	AD-5W (36), AD-4Q (2)
Sep 56	AD-5W (45), AD-4Q (2)
Oct 56	AD-5W (45)
Nov 56	AD-5W (38)
Dec 56	AD-5W (40)
Jan 57	AD-5W (45)
Feb 57	AD-5W (39)
Mar 57	AD-5W (38)
Apr 57	AD-5W (37)
May 57	AD-5W (36)
Jun 57	AD-5W (36)
Jul 57	AD-5W (35)
Aug 57	AD-5W (34)
Sep 57	AD-5W (36)
Oct 57	AD-5W (36)
Nov 57	AD-5W (46)
Dec 57	AD-5W (44)
Jan 58	AD-5W (49) CarQuals on Hornet
Feb 58	AD-5W (46)
Mar 58	AD-5W (44)
Apr 58	AD-5W (40)
May 58	AD-5W (30)
Jun 58	AD-5W (33)
Jul 58	AD-5W (30)
Aug 58	AD-5W (27)
Sep 58	AD-5W (28), F2H-4 (5)
Oct 58	AD-5W (32), F2H-4 (10)
Nov 58	AD-5W (31), F2H-4 (10), AD-5 (1)
Dec 58	AD-5W (34), F2H-4 (10), AD-5Q (1)
Jan 59	AD-5W (41), F2H-4 (10), AD-5Q (9)
Feb 59	AD-5W (38), F2H-4 (9), AD-5Q (14)
Mar 59	AD-5W (38), F2H-4 (7), AD-5Q (22)
Apr 59	AD-5W (39), F2H-4 (7), AD-5Q (23)
May 59	AD-5W (39), AD-5Q (24)
Jun 59	AD-5W (41), AD-5Q (23)
Jul 59	AD-5W (39), AD-5Q (24)
Aug 59	AD-5W (36), AD-5Q (20), AD-5 (1), TF-1Q (2), S2F-1 (1)

Sep 59	AD-5W (36), AD-5Q (20), AD-5 (1), TF-1Q (2), S2F-1 (1)
Oct 59	AD-5W (31), AD-5Q (19), AD-5 (1), TF-1Q (2), S2F-1 (1)
Nov 59	AD-5W (31), AD-5Q (17), AD-5 (1), TF-1Q (2), S2F-1 (1)
Dec 59	AD-5W (16), AD-5Q (13), AD-5 (1), TF-1Q (2), S2F-1 (1), S2F-2 (1), WF-2 (2)
Jan 60	AD-5W (16), AD-5Q (11), AD-5 (1), TF-1Q (2), S2F-2 (2), WF-2 (8)
Feb 60	AD-5W (16), AD-5Q (11), AD-5 (1), TF-1Q (2), S2F-2 (2), WF-2 (8)
Mar 60	AD-5W (16), AD-5Q (13), AD-5 (1), TF-1Q (2), S2F-2 (2), WF-2 (8)
Apr 60	AD-5W (17), AD-5Q (13), AD-5 (1), TF-1Q (2), S2F-2 (2), WF-2 (8)
May 60	AD-5W (17), AD-5Q (13), AD-5 (7), TF-1Q (2), S2F-2 (2), WF-2 (8)
Jun 60	AD-5W (17), AD-5Q (11), AD-5 (1), TF-1Q (2), S2F-2 (2), WF-2 (8)
Jul 60	AD-5W (13), AD-5Q (10), TF-1Q (2), S2F-2 (2), WF-2 (14)
Aug 60	AD-5W (18), AD-5Q (9), TF-1Q (2), S2F-2 (2), WF-2 (15)
Sep 60	AD-5W (19), AD-5Q (10), TF-1Q (2), S2F-2 (2), WF-2 (17)
Oct 60	AD-5W (19), AD-5Q (9), TF-1Q (2), S2F-2 (2), WF-2 (16)
Nov 60	AD-5W (16), AD-5Q (8), TF-1Q (2), S2F-2 (2), WF-2 (17)
Dec 60	AD-5W (17), AD-5Q (8), TF-1Q (2), S2F-2 (1), WF-2 (19)
Jan 61	AD-5W (13), AD-5Q (8), TF-1Q (2), WF-2 (18)
Feb 61	AD-5W (10), AD-5Q (8), TF-1Q (2), WF-2 (18)
Mar 61	AD-5W (10), AD-5Q (6), TF-1Q (2), WF-2 (13)
Apr 61	AD-5W (11), AD-5Q (6), TF-1Q (1), WF-2 (11)
May 61	AD-5W (9), AD-5Q (6), TF-1Q (1), WF-2 (14)
Jun 61	AD-5W (11), AD-5Q (5), TF-1Q (1), WF-2 (15)
Jul 61	AD-5W (11), AD-5Q (1), WF-2 (11)
Aug 61	AD-5W (11), WF-2 (17)
Sep 61	AD-5W (11), WF-2 (20)
Oct 61	AD-5W (11), WF-2 (23)
Nov 61	AD-5W (13), WF-2 (15)
Dec 61	AD-5W (11), WF-2 (19)
Jan 62	AD-5W (9), WF-2 (19)
Feb 62	AD-5W (10), WF-2 (19)
Mar 62	AD-5W (13), WF-2 (21)
Apr 62	AD-5W (14), WF-2 (22)
May 62	AD-5W (14), WF-2 (25)
Jun 62	AD-5W (16), WF-2 (25)
Jul 62	AD-5W (12), WF-2 (21)
Aug 62	AD-5W (13), WF-2 (21)
Sep 62	AD-5W (16), WF-2 (23)
Oct 62	A-1E (2), EA-1E (17), E-1B (22)
Nov 62	A-1E (2), EA-1E (11), E-1B (16)
Dec 62	A-1E (2), EA-1E (16), E-1B (22)
Jan 63	A-1E (2), EA-1E (13), E-1B (16)
Feb 63	A-1E (2), EA-1E (12), E-1B (16)
Mar 63	A-1E (2), EA-1E (12), E-1B (11)
Apr 63	A-1E (2), EA-1E (5), E-1B (17)
May 63	A-1E (2), EA-1E (7), E-1B (17)
Jun 63	A-1E (2), EA-1E (9), E-1B (17)
Jul 63	A-1E (2), EA-1E (10), E-1B (13)
Aug 63	A-1E (2), EA-1E (9), E-1B (15)
Sep 63	A-1E (2), EA-1E (4), E-1B (20)
Oct 63	A-1E (2), EA-1E (3), E-1B (17)
Nov 63	A-1E (2), EA-1E (2), E-1B (17)
Dec 63	A-1E (2), EA-1E (6), E-1B (21)
Jan 64	A-1E (2), EA-1E (5), E-1B (17), E-2A (1)
Feb 64	A-1E (2), EA-1E (2), E-1B (15), E-2A (2)
Mar 64	A-1E (2), EA-1F (2), E-1B (20), E-2A (4)
Apr 64	A-1E (2), EA-1E (6), E-1B (21), E-2A (7)
May 64	A-1E (2), EA-1E (5), E-1B (8), E-2A (9)
Jun 64	A-1E (2), EA-1E (6), E-1B (18), E-2A (10)
Jul 64	A-1E (2), EA-1E (1), E-1B (21), E-2A (12)
Aug 64	A-1E (2), EA-1E (5), E-1B (16), E-2A (15)
Sep 64	A-1E (2), EA-1F (4), E-1B (11), E-2A (16)
Oct 64	A-1E (2), EA-1E (1), E-1B (8), E-2A (18)
Nov 64	A-1E (2), EA-1E (4), E-1B (9), E-2A (18)
Dec 64	A-1E (2), EA-1E (4), E-1B (9), E-2A (18)
Jan 65	A-1E (2), EA-1E (4), E-1B (11), E-2A (18)
Feb 65	A-1E (2), EA-1E (4), E-1B (11), E-2A (18)

1952

1954-55

1955-56

VC-11/VAW-11 DETS 1954-1965

Jan 54	Unit B	CVA-9	AD-4W (3)
	Unit D	CVA-15	AD-4W (3)
	Det F	CVA-33	AD-4W (3)
	Unit G	CVA-34	AD-4W (2)
	Det I	CVA-9	AD-4W (3)
	Unit M	CVA-47	AD-4W (3)
Feb 54	Unit A	CVA-10	AD-4W (1)
	Unit E	CVA-34	AD-4W (3)
	Det I	CVA-9	AD-4W (3)
	Det 36	CVA-15	AD-4W (3)
Mar 54	Unit B	CVA-47	AD-4W (3)
	Unit E	CVA-34	AD-4W (1)
	Det G	CVA-21	AD-4W (3)
	Det I	CVA-9	AD-4W (3)
	Det 36	CVA-15	AD-4W (3)
Apr 54	Unit B	CVA-47	AD-4W (3)
	Det G	CVA-21	AD-4W (3)
	Det I	CVA-9	AD-4W (3)
	Unit M	S.D.	AD-4W (3)
	Det 36	CVA-15	AD-4W (3)
May 54	Unit B	CVA-47	AD-4W (3)
	Det G	CVA-21	AD-4W (3)
	Det I	CVA-9	AD-4W (3)
	Unit M	S.D.	AD-4W (3)
	Det 36	CVA-15	AD-4W (3)
Jun 54	Unit B	CVA-47	AD-4W (3)
	Unit D	CVA-10	AD-4W (3)
	Det G	CVA-21	AD-4W (3)
	Det I	CVA-9	AD-4W (3)
	Unit M	CVA-12	AD-4W (3)
	Det 36	CVA-15	AD-4W (3)
Jul 54	Unit B	CVA-47	AD-4W (3)
	Unit D	CVA-10	AD-4W (3)
	Det G	CVA-21	AD-4W (3)
	Unit M	CVA-12	AD-4W (3)
	Det 36	CVA-15	AD-4W (3)
Aug 54	Unit B	CVA-47	AD-4W (3)
	Unit D	CVA-10	AD-4W (3)
	Det G	CVA-21	AD-4W (3)
	Unit M	CVA-12	AD-4W (3)
Sep 54	Unit B	CVA-47	AD-4W (3)
	Unit D	CVA-10	AD-4W (3)
	Det G	CVA-21	AD-4W (3)
	Det H	CVA-18	AD-4W (3)
	Unit M	CVA-12	AD-4W (3)
Oct 54	Unit B	CVA-47	AD-4W (3)
	Det C	CVA-33	AD-4W (3)
	Det D	CVA-10	AD-4W (3)
	Det H	CVA-18	AD-4W (3)
	Det M	CVA-12	AD-4W (3)
Nov 54	Det A	CVA-9	AD-4W (3)
	Det C	CVA-33	AD-4W (3)
	Det D	CVA-10	AD-4W (3)
	Det H	CVA-18	AD-4W (3)
	Det M	CVA-12	AD-4W (3)
Dec 54	Det A	CVA-9	AD-4W (3)
	Det C	CVA-33	AD-4W (3)
	Det D	CVA-10	AD-4W (3)
	Det H	CVA-18	AD-4W (3)
Jan 55	Det A	CVA-9	AD-4W (3)
	Det C	CVA-33	AD-4W (0)
	Det D	CVA-10	AD-4W (3)
	Det H	CVA-18	AD-4W (3)
Feb 55	Det A	CVA-9	AD-4W (3)
	Det C	CVA-33	AD-4W (4)
	Det E	CVA-34	AD-4W (3)
	Det H	CVA-18	AD-4W (3)
Mar 55	Det A	CVA-9	AD-4W (3)
	Det C	CVA-33	AD-4W (3)
	Det E	CVA-34	AD-4W (3)
	Det H	CVA-18	AD-4W (2)
	Det I	CVA-47	AD-4W (3)
Apr 55	Det A	CVA-9	AD-4W (3)
	Det C	CVA-33	AD-4W (3)

	Det E	CVA-34	AD-4W (3)
	Det I	CVA-47	AD-4W (3)
May 55	Det A	CVA-9	AD-4W (3)
	Det C	CVA-33	AD-4W (3)
	Det E	CVA-34	AD-4W (3)
	Det I	CVA-47	AD-4W (3)
Jun 55	Det E	CVA-34	AD-4W (3)
	Det F	CVA-21	AD-4W (4), AD-4Q (1)
	Det I	CVA-47	AD-4W (3)
Jul 55	Det E	CVA-34	AD-4W (3)
	Det F	CVA-21	AD-4W (3), AD-4Q (1)
	Det I	CVA-47	AD-4W (3)
Aug 55	Det E	CVA-34	AD-4W (3)
	Det F	CVA-21	AD-4W (3), AD-4Q (1)
	Det G	CVA-19	AD-5W (3), AD-4Q (1)
	Det I	CVA-47	AD-4W (3)
Sep 55	Det F	CVA-21	AD-4W (3), AD-4Q (1)
	Det G	CVA-19	AD-5W (3), AD-4Q (1)
	Det I	CVA-47	AD-4W (3)
Oct 55	Det B	CVA-33	AD-5W (3), AD-4Q (1)
	Det F	CVA-21	AD-4W (3), AD-4Q (1)
	Det G	CVA-19	AD-5W (3), AD-4Q (1)
	Det I	CVA-47	AD-4W (3)
Nov 55	Det B	CVA-33	AD-5W (3), AD-4Q (1)
	Det F	CVA-21	AD-4W (3), AD-4Q (1)
	Det G	CVA-19	AD-5W (2), AD-4Q (1)
Dec 55	Det B	CVA-33	AD-5W (2), AD-4Q (1)
	Det F	CVA-21	AD-4W (3)
	Det G	CVA-19	AD-5W (2), AD-4Q (1)
Jan 56	Det B	CVA-33	AD-5W (3), AD-4Q (1)
	Det F	CVA-21	AD-4W (3), AD-4Q (1)
	Det G	CVA-19	AD-5W (2), AD-4Q (1)
	Det J	CVA-38	AD-5W (3), AD-4Q (1)
	Det M	CVA-34	AD-5W (3), AD-4Q (1)
Feb 56	Det B	CVA-33	AD-5W (3), AD-4Q (1)
	Det G	CVA-19	AD-5W (2), AD-4Q (1)
	Det J	CVA-38	AD-5W (3), AD-4Q (1)
	Det M	CVA-34	AD-5W (3), AD-4Q (1)
Mar 56	Det B	CVA-33	AD-5W (3), AD-4Q (1)
	Det J	CVA-38	AD-5W (3), AD-4Q (1)
	Det K	CVA-10	AD-5W (3), AD-4Q (1)
	Det M	CVA-34	AD-5W (3), AD-4Q (1)
Apr 56	Det B	CVA-33	AD-5W (3), AD-4Q (1)
	Det D	CVA-18	AD-5W (3), AD-4Q (1)
	Det J	CVA-38	AD-5W (3), AD-4Q (1)
	Det K	CVA-10	AD-5W (3), AD-4Q (1)
	Det M	CVA-34	AD-5W (3), AD-4Q (1)
May 56	Det D	CVA-18	AD-5W (3), AD-4Q (1)
	Det H	CVA-16	AD-5W (3)
	Det J	CVA-38	AD-5W (3), AD-4Q (1)
	Det K	CVA-10	AD-5W (3), AD-4Q (1)
	Det M	CVA-34	AD-5W (3), AD-4Q (1)
Jun 56	Det D	CVA-18	AD-5W (3), AD-4Q (1)
	Det H	CVA-16	AD-5W (3)
	Det K	CVA-10	AD-5W (3), AD-4Q (1)
	Det M	CVA-34	AD-5W (3), AD-4Q (1)

VAW-11

Jul 56	Det C	CVA-9	AD-5W (3)
	Det D	CVA-18	AD-5W (3), AD-4Q (1)
	Det H	CVA-16	AD-5W (3)
	Det K	CVA-10	AD-5W (3), AD-4Q (1)
	Det M	CVA-34	AD-5W (3), AD-4Q (1)
Aug 56	Det C	CVA-9	AD-5W (3)
	Det D	CVA-18	AD-5W (3), AD-4Q (1)
	Det H	CVA-16	AD-5W (3)
	Det K	CVA-10	AD-5W (3), AD-4Q (1)
	Det L	CVA-31	AD-5W (3)
Sep 56	Det C	CVA-9	AD-5W (3)
	Det D	CVA-18	AD-5W (3), AD-4Q (1)
	Det H	CVA-16	AD-5W (3)
	Det L	CVA-31	AD-5W (3)
Oct 56	Det C	CVA-9	AD-5W (3)
	Det D	CVA-18	AD-5W (3), AD-4Q (1)
	Det H	CVA-16	AD-5W (3)
	Det L	CVA-31	AD-5W (3)

Above, VAW-11 AD-5Q BuNo 132589 at NAS North Island, CA, on 18 March 1960. (William Swisher) At right, VAW-11 EA-1E BuNo 132751 taxiing on the USS Kearsarge (CVS-33) in 1958. (Paul Minert collection) Below, VAW-11 AD-5W BuNo 135143 taxing at NAS North Island, CA, in 1960. (USN)

tail code "RR", illustrating the need for such a large number of AD-5Ws and AD-5Qs. Another five carriers would be fielded between 1959 and the end of 1969.

In August 1959, the squadron received two TF-1Qs and one S2F-1. In December 1959, two WF-2s were

DET H

DET I

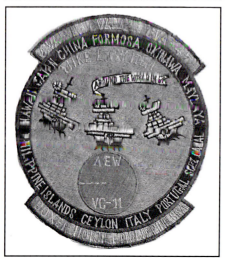

Date	Det	Ship/Base	Aircraft
Nov 56	Det A	CVA-38	AD-5W (3)
	Det C	CVA-9	AD-5W (3)
	Det E	CVA-10	AD-5W (3)
	Det H	CVA-16	AD-5W (3)
	Det L	CVA-31	AD-5W (3)
Dec 56	Det A	CVA-38	AD-5W (3)
	Det C	CVA-9	AD-5W (3)
	Det E	CVA-10	AD-5W (3)
	Det F	CVA-12	AD-5W (3)
	Det L	CVA-31	AD-5W (3)
Jan 57	Det A	CVA-38	AD-5W (3)
	Det E	CVA-10	AD-5W (3)
	Det F	CVA-12	AD-5W (3)
	Det L	CVA-31	AD-5W (3)
Feb 57	Det A	CVA-38	AD-5W (3)
	Det E	CVA-10	AD-5W (3)
	Det F	CVA-12	AD-5W (3)
Mar 57	Det A	CVA-38	AD-5W (3)
	Det E	CVA-10	AD-5W (3)
	Det F	CVA-12	AD-5W (3)
Apr 57	Det A	CVA-38	AD-5W (3)
	Det E	CVA-10	AD-5W (3)
	Det F	CVA-12	AD-5W (3)
	Det G	CVA-16	AD-5W (3)
	Det I	CVA-19	AD-5W (2)
May 57	Det E	CVA-10	AD-5W (3)
	Det F	CVA-12	AD-5W (3)
	Det G	CVA-16	AD-5W (3)
	Det I	CVA-19	AD-5W (2)
Jun 57	Det E	CVA-10	AD-5W (3)
	Det F	CVA-12	AD-5W (3)
	Det G	CVA-16	AD-5W (3)
	Det I	CVA-19	AD-5W (3)
Jul 57	Det B	CVA-31	AD-5W (3)
	Det E	CVA-10	AD-5W (3)
	Det G	CVA-16	AD-5W (3)
	Det I	CVA-19	AD-5W (3)
Aug 57	Det B	CVA-31	AD-5W (3)
	Det G	CVA-16	AD-5W (3)
	Det I	CVA-19	AD-5W (3)
	Det J	CVA-33	AD-5W (3)
Sep 57	Det B	CVA-31	AD-5W (3)
	Det G	CVA-16	AD-5W (2)
	Det J	CVA-33	AD-5W (3)
	Det M	CVA-14	AD-5W (3)
Oct 57	Det B	CVA-31	AD-5W (3)
	Det J	CVA-33	AD-5W (3)
	Det M	CVA-14	AD-5W (3)
Nov 57	Det B	CVA-31	AD-5W (3)
	Det J	CVA-33	AD-5W (3)
	Det M	CVA-14	AD-5W (3)
Dec 57	Det D	Noris	AD-5W (3)
	Det J	CVA-33	AD-5W (3)
	Det K	Noris	AD-5W (3)
	Det M	CVA-14	AD-5W (3)
Jan 58	Det D	Noris	AD-5W (3)
	Det J	CVA-33	AD-5W (3)
	Det K	Noris	AD-5W (3)
	Det M	CVA-14	AD-5W (3)
Feb 58	Det C	Noris	AD-5W (3)
	Det D	CVA-19	AD-5W (3)
	Det J	CVA-33	AD-5W (3)
	Det K	Atsugi	AD-5W (3)
	Det M	CVA-14	AD-5W (3)
Mar 58	Det C	CVA-38	AD-5W (3)
	Det D	Atsugi	AD-5W (3)
	Det J	CVA-33	AD-5W (3)
	Det K	Atsugi	AD-5W (3)
	Det M	CVA-14	AD-5W (3)
Apr 58	Det C	Agana	AD-5W (4)
	Det D	CVA-19	AD-5W (3)
	Det H	Noris	AD-5W (3)
	Det J	CVA-33	AD-5W (3)
	Det K	Agana	AD-5W (3)
May 58	Det C	Agana	AD-5W (3), AD-5 (1)
	Det D	Cubi	AD-5W (3)
	Det F	CVA-61	AD-5W (3)
	Det H	CVA-20	AD-5W (3)
	Det K	Agana	AD-5W (3)
Jun 58	Det A	Noris	AD-5W (3)
	Det C	Agana	AD-5W (3), AD-5 (1)
	Det D	Cubi	AD-5W (3)
	Det F	CVA-61	AD-5W (3)
	Det H	CVA-20	AD-5W (3)
	Det K	CVA-12	AD-5W (2)
	Det L	Noris	AD-5W (3)
Jul 58	Det A	CVA-41	AD-5W (3)
	Det C	Agana	AD-5W (3), AD-5 (1)
	Det D	Cubi	AD-5W (3)
	Det F	CVA-61	AD-5W (3)
	Det H	CVA-14	AD-5W (3)
	Det L	CVA-16	AD-5W (3)
Aug 58	Det A	CVA-41	AD-5W (2)
	Det C	Atsugi	AD-5W (3), AD-5 (1)
	Det D	Atsugi	AD-5W (3)
	Det E	Noris	AD-5W (3)
	Det H	Alameda	AD-5W (3)
	Det K	Noris	AD-5W (3)
	Det L	CVA-16	AD-5W (3)
Sep 58	Det A	CVA-41	AD-5W (2)
	Det C	Atsugi	AD-5W (4)
	Det D	CVA-19	AD-5W (3), AD-5 (1)
	Det E	Noris	AD-5W (3)
	Det H	CVA-14	AD-5W (3)
	Det K	Noris	AD-5W (3)
	Det L	CVA-16	AD-5W (3)
Oct 58	Det A	CVA-41	AD-5W (2)
	Det C	Atsugi	AD-5 (4)
	Det E	Noris	AD-5W (3)
	Det H	CVA-14	AD-5W (3)
	Det K	Noris	AD-5W (3)
	Det L	CVA-16	AD-5W (3)
Nov 58	Det A	CVA-41	AD-5W (2)
	Det E	Noris	AD-5W (3)
	Det F	Noris	AD-5W (3)
	Det H	Cubi	AD-5W (3)
	Det K	Noris	AD-5W (3)
	Det L	CVA-16	AD-5W (3)
Dec 58	Det A	CVA-41	AD-5W (2)
	Det E	Cubi	AD-5W (3), AD-5 (1)
	Det F	CVA-61	AD-5W (3)
	Det H	Cubi	AD-5W (3)
	Det K	Noris	AD-5W (3)
Jan 59	Det A	CVA-41	AD-5W (2)
	Det E	Cubi	AD-5W (3), AD-5 (1)
	Det F	CVA-61	AD-5W (3)
	Det H	CVA-14	AD-5W (3)
Feb 59	Det A	CVA-41	AD-5W (2)
	Det C	Noris	AD-5W (3)
	Det E	CVA-31	AD-5W (3), AD-5 (1)
	Det F	Atsugi	AD-5W (3), AD-5 (1)
Mar 59	Det C	Bar. Pt.	AD-5W (3)
	Det E	CVA-31	AD-5W (3), AD-5 (1)
	Det F	Atsugi	AD-5W (3), AD-5 (1)
	Det P	Noris	F2H-4 (5)
Apr 59	Det C	Bar. Pt.	AD-5W (3)
	Det E	CVA-31	AD-5W (3), AD-5 (1)
	Det F	Cubi	AD-5W (3), AD-5 (1)
	Det L	CVA-16	AD-5W (3)
	Det P	CVA-12	F2H-4 (4)
May 59	Det C	Bar. Pt.	AD-5W (3)
	Det E	CVA-31	AD-5W (3), AD-5 (1)
	Det FW	CVA-61	AD-5W (3), AD-5 (1)
	Det FN	CVA-61	AD-5N (1)
	Det LW	CVA-16	AD-5W (3)
	Det LN	CVA-16	AD-5N (3)
	Det P	CVA-12	F2H-4 (5)
Jun 59	Det C	Bar. Pt.	AD-5W (3)
	Det E	Noris	AD-5 (1)
	Det FW	CVA-61	AD-5W (2), AD-5 (1)

Above, VAW-11 AD-5W begins launch from the USS Midway (CVA-41) on 13 September 1956. Tail stripe was maroon. (USN) At right, VAW-11 AD-5W BuNo 135212 trapping during SEATO Exercise "Sea Lion" in the South China Sea on 5 April 1960. (Hal Andrews via Jim Sullivan) Below, VAW-11 AD-5W BuNo 132784 flies past Mt. Fuji, Japan, while assigned to ATG-4. (Ginter collection)

acquired and the slow replacement of the Skyraiders began. However, the WF-2/E-1B never fully replaced the AD-5W/EA-1E as the E-2A arrived in

1962   CVS-10

	Det FN	CVA-61	AD-5N (1)
	Det LW	CVA-16	AD-5W (3)
	Det LN	CVA-16	AD-5N (2)
	Det P	CVA-12	F2H-4 (4), F2H-3 (1)
Jul 59	Det A	CVA-41	AD-5W (3)
	Det C	Bar. Pt.	AD-5W (3)
	Det CN	CVA-38	AD-5N (2)
	Det D	CVA-19	AD-5W (3)
	Det LW	CVA-16	AD-5W (3)
	Det LN	CVA-16	AD-5N (2)
	Det P	CVA-12	F2H-4 (4), F2H-3 (1)
Aug 59	Det A	CVA-41	AD-5W (3)
	Det C	Bar. Pt.	AD-5W (3)
	Det CN	CVA-38	AD-5N (2)
	Det D	CVA-19	AD-5W (3)
	Det LW	CVA-16	AD-5W (3)
	Det LN	CVA-16	AD-5N (2)
	Det P	CVA-12	F2H-4 (4), F2H-3 (1)
Sep 59	Det C	Bar. Pt.	AD-5W (3)
	Det CN	CVA-38	AD-5N (2)
	Det LW	CVA-16	AD-5W (3)
	Det LN	CVA-16	AD-5N (2)
	Det P	CVA-12	F2H-4 (4), F2H-3 (1)
Oct 59	Det LW	CVA-16	AD-5W (3)
	Det LN	CVA-16	AD-5N (2)
Nov 59	Det E	CVA-31	AD-5Q (2)
	Det LW	CVA-16	AD-5W (1)
	Det LN	CVA-16	AD-5N (1)
Dec 59	Det H	Noris	AD-5W (3), AD-5Q (3)
	Det T	Noris	AD-5W (4)
Jan 60	Det H	Noris	AD-5W (3), AD-5Q (3)
	Det T	CVS-10	AD-5W (4)
Feb 60	Det M	Noris	AD-5W (3), AD-5Q (4)
	Det T	CVS-10	AD-5W (4)
Mar 60	Det T	CVS-10	AD-5W (4)
Apr 60	Det T	CVS-10	AD-5W (4)
May 60	Det F	Bar. Pt.	AD-5Q (2)
	Det T	CVS-10	AD-5W (4)
Jun 60	Det C	Bar. Pt.	WF-2 (3)
	Det T	CVS-10	AD-5W (4)
Jul 60	Det C	CVA-19	WF-2 (3)
	Det Q	Noris	AD-5W (4)
	Det T	Noris	AD-5W (4)
Aug 60	Det Q	Noris	AD-5W (4)
Sep 60	Det D	CVA-43	WF-2 (3)
	Det Q	CVS-20	AD-5W (4)
Oct 60	Det L	CVA-16	WF-2 (3)
	Det Q	CVS-20	AD-5W (4)
Nov 60	Det Q	CVS-20	AD-5W (4)
Dec 60	Det Q	CVS-20	AD-5W (4)
Jan 61	Det A	CVA-41	WF-2 (3)
	Det Q	CVS-20	AD-5W (4)
	Det R	CVS-33	AD-5W (4)
Feb 61	Det A	CVA-41	WF-2 (3)
	Det Q	CVS-20	AD-5W (4)
	Det R	CVS-33	AD-5W (6)
Mar 61	Det A	CVA-41	WF-2 (3)
	Det E	Noris	WF-2 (3), AD-5Q (2)
	Det Q	CVS-20	AD-5W (4)
	Det R	CVS-33	AD-5W (4)
Apr 61	Det A	CVA-41	WF-2 (3)
	Det E	CVA-31	WF-2 (3), AD-5Q (2)
	Det Q	CVS-20	AD-5W (4)
	Det R	CVS-33	AD-5W (4)
May 61	Det A	Atsugi	WF-2 (3)
	Det E	CVA-31	WF-2 (3), AD-5Q (2)
	Det R	CVS-33	AD-5W (6)
	Det T	Noris	AD-5W (4)
Jun 61	Det A	Atsugi	WF-2 (3)
	Det B	CVA-14	WF-2 (3)
	Det E	CVA-31	WF-2 (3), AD-5Q (2)
	Det M	CVA-61	WF-2 (3), AD-5Q (2)
	Det R	CVS-33	AD-5W (5)
	Det T	Noris	AD-5W (4)
Jul 61	Det A	Atsugi	WF-2 (3)

	Det B	CVA-14	WF-2 (3)
	Det E	CVA-31	WF-2 (3)
	Det M	Miramar	WF-2 (3), AD-5Q (2)
	Det R	CVS-33	AD-5W (5)
	Det T	CVS-10	AD-5W (4)
Aug 61	Det A	Atsugi	WF-2 (3)
	Det B	CVA-14	WF-2 (3)
	Det E	CVA-31	WF-2 (3)
	Det M	CVA-61	WF-2 (5)
	Det R	CVS-33	AD-5W (5)
	Det T	CVS-10	AD-5W (4)
Sep 61	Det B	CVA-14	WF-2 (3)
	Det E	CVA-31	WF-2 (3)
	Det F	Noris	WF-2 (3)
	Det M	CVA-61	WF-2 (4)
	Det Q	Noris	AD-5W (4)
	Det T	CVS-10	AD-5W (4)
	Det I	Atsugi	WF-2 (1)
Oct 61	Det B	CVA-14	WF-2 (3)
	Det E	CVA-31	WF-2 (3)
	Det F	Noris	WF-2 (3)
	Det M	CVA-61	WF-2 (4)
	Det Q	Noris	AD-5W (4)
	Det T	CVS-10	AD-5W (4)
	Det I	Atsugi	WF-2 (1)
Nov 61	Det B	CVA-14	WF-2 (3)
	Det D	Noris	WF-2 (4)
	Det E	CVA-31	WF-2 (3)
	Det F	CVA-16	WF-2 (3)
	Det L	Moffett	WF-2 (3)
	Det M	CVA-61	WF-2 (4)
	Det Q	Noris	AD-5W (4)
	Det T	CVS-10	AD-5W (4)
	Det I	Atsugi	WF-2 (1)
Dec 61	Det B	CVA-14	WF-2 (3)
	Det D	Noris	WF-2 (4)
	Det F	CVA-16	WF-2 (3)
	Det L	Noris	WF-2 (4)
	Det M	CVA-61	WF-2 (4)
	Det Q	Noris	AD-5W (5)
	Det T	CVS-10	AD-5W (3)
Jan 62	Det A	Noris	WF-2 (4)
	Det D	CVA-43	WF-2 (4)
	Det F	CVA-16	WF-2 (3)
	Det L	CVA-19	WF-2 (3)
	Det M	CVA-61	WF-2 (4)
	Det N	Noris	AD-5W (4)
	Det Q	CVS-20	AD-5W (5)
	Det T	CVS-10	AD-5W (3)
Feb 62	Det A	Noris	WF-2 (4)
	Det D	CVA-43	WF-2 (4)
	Det F	CVA-16	WF-2 (3)
	Det L	CVA-19	WF-2 (3)
	Det M	CVA-61	WF-2 (4)
	Det N	Noris	AD-5W (4)
	Det Q	CVS-20	AD-5W (5)
	Det T	CVS-10	AD-5W (3)
Mar 62	Det A	Noris	WF-2 (4)
	Det D	CVA-43	WF-2 (4)
	Det F	CVA-16	WF-2 (3)
	Det G	Noris	WF-2 (3)
	Det L	CVA-19	WF-2 (2)
	Det N	Noris	AD-5W (5)
	Det Q	CVS-20	AD-5W (5)
Apr 62	Det A	CVA-41	WF-2 (4)
	Det D	CVA-43	WF-2 (4)
	Det F	CVA-16	WF-2 (3)
	Det G	Noris	WF-2 (3)
	Det L	CVA-19	WF-2 (3)
	Det N	Noris	AD-5W (5)
	Det Q	CVS-20	AD-5W (5)
May 62	Det A	CVA-41	WF-2 (4)
	Det D	CVA-43	WF-2 (4)
	Det G	Noris	WF-2 (3)
	Det L	CVA-19	WF-2 (3)

Date	Det	Unit	Aircraft
	Det N	Noris	AD-5W (5)
	Det Q	CVS-20	AD-5W (5)
Jun 62	Det A	CVA-41	WF-2 (4)
	Det D	CVA-43	WF-2 (4)
	Det G	CVA-34	WF-2 (2)
	Det L	CVA-19	WF-2 (3)
	Det M	Noris	AD-5W (5)
	Det N	CVS-12	AD-5W (5)
	Det Q	CVS-20	AD-5W (5)
	Det I	Cubi	AD-5Q (2)
Jul 62	Det A	CVA-41	WF-2 (4)
	Det C	Noris	WF-2 (2)
	Det G	CVA-34	WF-2 (3)
	Det L	CVA-19	WF-2 (3)
	Det M	Noris	WF-2 (4)
	Det N	CVS-12	AD-5W (5)
	Det Q	Noris	AD-5W (4)
	Det R	Noris	AD-5W (3)
Aug 62	Det A	CVA-41	WF-2 (4)
	Det C	Noris	WF-2 (2)
	Det G	CVA-34	WF-2 (2)
	Det M	Noris	WF-2 (4)
	Det N	CVS-12	AD-5W (5)
	Det R	Ford Is.	AD-5W (3)
Sep 62	Det A	CVA-41	WF-2 (4)
	Det C	CVA-63	WF-2 (3)
	Det G	CVA-34	WF-2 (3)
	Det M	Noris	WF-2 (4)
	Det N	CVS-12	AD-5W (5)
	Det R	CVS-33	AD-5W (3)
Oct 62	Det B	CVA-14	E-1B (3)
	Det C	CVA-63	E-1B (3)
	Det E	CVA-31	E-1B (1)
	Det G	CVA-34	E-1B (3)
	Det M	Noris	E-1B (4)
	Det N	CVS-12	EA-1E (5)
Nov 62	Det B	CVA-14	E-1B (3)
	Det C	CVA-63	E-1B (3)
	Det E	CVA-31	E-1B (1)
	Det F	CVA-64	E-1B (4)
	Det G	CVA-34	E-1B (3)
	Det M	CVA-61	E-1B (4)
	Det N	CVS-12	EA-1E (5)
	Det Q	Noris	EA-1E (5)
Dec 62	Det B	CVA-14	E-1B (3)
	Det C	CVA-63	E-1B (4)
	Det E	CVA-31	E-1B (1)
	Det F	Noris	E-1B (4)
	Det M	CVA-61	E-1B (4)
	Det Q	Noris	EA-1E (5)
	Det T	CVS-10	EA-1E (1)
Jan 63	Det B	CVA-14	E-1B (3)
	Det C	CVA-63	E-1B (4)
	Det D	Noris	E-1B (4)
	Det E	CVA-31	E-1B (1)
	Det F	Noris	E-1B (4)
	Det M	CVA-61	E-1B (4)
	Det Q	Noris	EA-1E (5)
	Det T	CVS-10	EA-1E (1)
Feb 63	Det B	CVA-14	E-1B (3)
	Det C	CVA-63	E-1B (4)
	Det D	Noris	E-1B (4)
	Det E	CVA-31	E-1B (1)
	Det F	CVA-64	E-1B (4)
	Det L	Miramar	E-1B (3)
	Det M	CVA-61	E-1B (3)
	Det Q	Noris	EA-1E (4)
	Det T	CVS-10	EA-1E (2)
Mar 63	Det B	CVA-14	E-1B (3)
	Det C	Noris	E-1B (4)
	Det D	Noris	E-1B (4)
	Det F	CVA-64	E-1B (4)
	Det G	Noris	E-1B (3)
	Det L	CVA-19	E-1B (3)
	Det M	CVA-61	E-1B (3)
	Det Q	Noris	EA-1E (3)
	Det R	Noris	EA-1E (1)
	Det T	CVS-10	EA-1E (2)
Apr 63	Det B	CVA-14	EA-1B (3)
	Det D	CVA-43	EA-1B (4)
	Det F	CVA-64	EA-1B (4)
	Det G	Noris	EA-1B (3)
	Det L	Noris	EA-1B (3)
	Det M	CVA-61	EA-1B (4)
	Det N	Noris	EA-1E (5)
	Det Q	Noris	EA-1E (3)
	Det R	CVA-34	EA-1E (2)
	Det T	CVS-10	EA-1E (2)
May 63	Det B	CVA-14	EA-1B (3)
	Det C	Noris	EA-1B (4)
	Det D	CVA-43	EA-1B (4)
	Det F	CVA-64	EA-1B (4)
	Det G	Noris	EA-1B (3)
	Det L	Noris	EA-1B (3)
	Det M	CVA-61	EA-1B (4)
	Det N	Noris	EA-1E (5)
	Det Q	Noris	EA-1E (3)
	Det R	CVS-37	EA-1E (2)
Jun 63	Det B	CVA-14	EA-1B (3)
	Det C	Noris	EA-1B (4)
	Det D	CVA-43	EA-1B (4)
	Det F	CVA-64	EA-1B (3)
	Det G	Noris	EA-1B (3)
	Det L	CVA-19	EA-1B (3)
	Det N	Noris	EA-1E (5)
	Det Q	Noris	EA-1E (3)
	Det R	CVS-37	EA-1E (2)
Jul 63	Det A	Noris	EA-1B (4)
	Det C	Noris	EA-1B (4)
	Det D	CVA-43	EA-1B (4)
	Det F	CVA-64	EA-1B (4)
	Det G	CVA-34	EA-1B (3)
	Det L	CVA-19	EA-1B (3)
	Det N	CVS-12	EA-1E (5)
	Det Q	Noris	EA-1E (3)
	Det R	CVS-34	EA-1E (2)
Aug 63	Det A	Noris	EA-1B (4)
	Det C	Noris	EA-1B (4)
	Det D	CVA-43	EA-1B (4)
	Det F	CVA-64	EA-1B (4)
	Det G	CVA-34	EA-1B (3)
	Det L	CVA-19	EA-1B (3)
	Det N	Noris	EA-1E (4)
	Det Q	Noris	EA-1E (3)
	Det R	CVS-33	EA-1E (2)
Sep 63	Det A	Noris	EA-1B (4)
	Det C	Noris	EA-1B (4)
	Det D	CVA-43	EA-1B (3)
	Det G	CVA-34	EA-1B (3)
	Det L	CVA-19	EA-1B (3)
	Det N	Noris	EA-1E (4)
	Det Q	Noris	EA-1E (8)
	Det R	CVS-33	EA-1E (2)
Oct 63	Det A	Noris	EA-1B (2)
	Det C	CVA-63	EA-1B (4)
	Det D	CVA-43	EA-1B (3)
	Det E	Noris	EA-1B (2)
	Det G	CVA-34	EA-1B (3)
	Det L	CVA-19	EA-1B (3)
	Det N	CVS-12	EA-1E (3)
	Det Q	Noris	EA-1E (7)
	Det R	CVS-33	EA-1E (1)
Nov 63	Det A	CVA-41	EA-1B (4)
	Det C	CVA-63	EA-1B (4)
	Det D	Noris	EA-1B (3)
	Det E	CVA-31	EA-1B (3)
	Det G	CVA-34	EA-1B (3)
	Det L	CVA-19	EA-1B (3)
	Det N	CVS-12	EA-1E (5)
	Det Q	CVS-20	EA-1E (8)
	Det R	CVS-33	EA-1E (1)
Dec 63	Det A	CVA-41	EA-1B (4)
	Det B	Noris	EA-1B (3)
	Det C	CVA-63	EA-1B (4)
	Det E	Noris	EA-1B (2)
	Det G	CVA-34	EA-1B (3)
	Det N	CVS-12	EA-1E (5)
	Det Q	Ream F.	EA-1E (5)
Jan 64	Det A	CVA-41	EA-1B (4)
	Det B	Noris	EA-1B (3)
	Det C	CVA-63	EA-1B (4)
	Det E	CVA-31	EA-1B (3)
	Det F	Noris	EA-1B (4)
	Det G	CVA-34	EA-1B (3)
	Det N	CVS-12	EA-1E (5)
	Det Q	Ream F.	EA-1E (5)
Feb 64	Det A	CVA-41	EA-1B (3)
	Det B	Noris	EA-1B (3)
	Det C	CVA-63	EA-1B (4)
	Det E	CVA-31	EA-1B (3)
	Det F	Noris	EA-1B (4)
	Det G	CVA-34	EA-1B (3)
	Det N	CVS-12	EA-1E (3)
	Det Q	CVS-20	EA-1E (5)
	Det R	Noris	EA-1E (2)
Mar 64	Det A	CVA-41	EA-1B (4)
	Det B	Noris	EA-1B (3)
	Det C	CVA-63	EA-1B (4)
	Det E	CVA-31	EA-1B (3)
	Det F	Noris	EA-1B (4)
	Det N	CVS-12	EA-1E (3)
	Det Q	CVS-20	EA-1E (5)
	Det R	Noris	EA-1E (3)
Apr 64	Det A	CVA-41	EA-1B (4)
	Det B	Noris	EA-1B (3)
	Det C	CVA-63	EA-1B (4)
	Det E	CVA-31	EA-1B (3)
	Det F	Noris	EA-1B (4)
	Det Q	CVS-20	EA-1E (5)
	Det R	CVS-33	EA-1E (4)
May 64	Det B	Noris	EA-1B (3)
	Det C	CVA-63	EA-1B (4)
	Det E	CVA-31	EA-1B (2)
	Det F	CVA-64	EA-1B (4)
	Det M	Noris	EA-1B (5)
	Det Q	CVS-20	EA-1E (5)
	Det R	Noris	EA-1E (5)
Jun 64	Det A	Noris	EA-1B (1)
	Det B	CVA-14	EA-1B (3)
	Det C	CVA-63	EA-1B (4)
	Det E	CVA-31	EA-1B (3)
	Det F	CVA-64	EA-1B (4)
	Det M	CVS-45	EA-1B (4)
	Det Q	CVS-20	EA-1E (5)
	Det R	CVS-33	EA-1E (5)
Jul 64	Det B	CVA-14	EA-1B (3)
	Det C	Noris	EA-1B (4)
	Det E	CVA-31	EA-1B (3)
	Det F	CVA-64	EA-1B (4)
	Det L	Noris	EA-1B (3)
	Det M	CVA-61	EA-1B (4)
	Det Q	CVS-20	EA-1E (5)
	Det R	CVS-33	EA-1E (5)
	Det T	Noris	EA-1E (1)
Aug 64	Det B	CVA-14	EA-1B (3)
	Det E	CVA-31	EA-1B (3)
	Det F	CVA-64	EA-1B (4)
	Det L	Noris	EA-1B (3)
	Det M	CVA-61	EA-1B (4)
	Det Q	Noris	EA-1E (5)
	Det R	CVS-33	EA-1E (5)
	Det T	Noris	EA-1E (5)
Sep 64	Det B	CVA-14	EA-1B (3)
	Det C	CVA-43	EA-1B (4)
	Det E	CVA-31	EA-1B (3)
	Det F	CVA-64	EA-1B (4)
	Det L	CVA-19	EA-1B (3)
	Det M	CVA-61	EA-1B (4)
	Det R	CVS-33	EA-1E (5)
	Det T	Noris	EA-1E (2)
Oct 64	Det B	CVA-14	EA-1B (3)
	Det D	Noris	EA-1B (4)
	Det E	CVA-31	EA-1B (3)
	Det F	CVA-64	EA-1B (4)
	Det G	Noris	EA-1B (3)
	Det L	CVA-19	EA-1B (3)
	Det R	CVS-33	EA-1E (5)
	Det T	CVS-10	EA-1E (5)
Nov 64	Det A	Noris	EA-1B (4)
	Det B	CVA-14	EA-1B (3)
	Det D	Noris	EA-1B (4)
	Det E	Noris	EA-1B (2)
	Det F	CVA-64	EA-1B (4)
	Det G	Noris	EA-1B (3)
	Det L	CVA-19	EA-1B (3)
	Det M	CVA-61	EA-1B (4)
	Det Q	Noris	EA-1B (4)
	Det R	CVS-33	EA-1E (5)
	Det T	CVS-10	EA-1E (5)
Dec 64	Det A	Noris	EA-1B (4)
	Det D	CVA-43	EA-1B (4)
	Det F	CVA-64	EA-1B (4)
	Det G	Noris	EA-1B (3)
	Det L	CVA-19	EA-1B (3)
	Det M	CVA-61	EA-1B (4)
	Det Q	Noris	EA-1B (4)
	Det T	CVS-10	EA-1E (5)
Jan 65	Det A	Noris	EA-1B (4)
	Det D	CVA-43	EA-1B (4)
	Det F	Noris	EA-1B (4)
	Det G	Noris	EA-1B (3)
	Det L	CVA-19	EA-1B (4)
	Det M	CVA-61	EA-1B (4)
	Det Q	Noris	EA-1B (4)
	Det T	CVS-10	EA-1E (1)
Feb 65	Det A	CVA-41	EA-1B (4)
	Det D	CVA-43	EA-1B (4)
	Det E	Noris	EA-1B (1)
	Det G	Noris	EA-1B (3)
	Det L	CVA-19	EA-1B (3)
	Det M	CVA-61	EA-1B (3)
	Det Q	Noris	EA-1B (4)
	Det T	CVS-10	EA-1E (1)
Mar 65	Det A	CVA-41	EA-1B (4)
	Det D	CVA-43	EA-1B (3)
	Det E	Noris	EA-1B (1)
	Det G	Noris	EA-1B (3)
	Det L	CVA-19	EA-1B (3)
	Det M	CVA-61	EA-1B (3)
	Det Q	CVS-20	EA-1B (4)
	Det T	CVS-10	EA-1E (5)

Above, VAW-11 EA-1E BuNo 139605 landing at NAS Atsugi, Japan, on 3 December 1962. Aircraft was assigned to the USS Yorktown (CVS-10) as part of CVSG-55. (Toyokazu Matsuzaki) Below, VAW-11 AD-5W BuNo 139565 (RR/703) landing at NAS Atsugi, Japan, on 24 July 1961. Aircraft was assigned to CVSG-53 aboard the USS Kearsarge (CVS-33) (Toyokazu Matsuzaki)

February 1964 to begin replacing them both. The last EA-1B Skyraider was retired from VAW-11 in April 1965.

Above, VAW-11 AD-5W landing on the USS Bennington (CVS-20) in 1962. (USN) Below, two VAW-11 AD-5Ws BuNos 139596 and 139568 on deck on CVA-31 with VA-54 AD-7s and a VAW-13 AD-5N. (USN)

COMPOSITE SQUADRON TWELVE, VC-12
AIRBORNE EARLY WARNING SQUADRON TWELVE, VAW-12

Carrier Airborne Early Warning Squadron Two (VAW-2) was established on 6 July 1948 with twelve TBM-3Ws, ten TBM-3Es, one SNJ and one SNB. With these assets the AEW Team/Detachments of three to four aircraft were dispatched to the Atlantic Fleet CVs. VAW-2 was redesignated Composite Squadron Twelve (VC-12) on 1 September 1948. While stationed at NAS Quonset Pt., RI, VC-12 received its first Skyraider, an AD-1Q in August 1948.

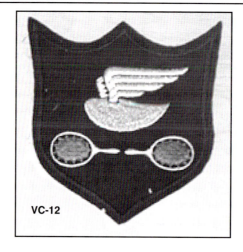

VC-12

They operated less than half a dozen AD-1/1Qs and AD-2s through July 1949 before adding two mission capable AD-3Ws and four AD-3s in August. By November 1949, the squadron had acquired twenty AD-3Ws and had conducted its first AD-3W carrier operations during a short five week cruise aboard the USS Wright (CVL-49). The squadron's first two AD-4Ws and four AD-4s were received in June 1950. The AD-5 arrived in August 1954 and the first AD-5W was acquired in November 1954. An AD-3Q was briefly used from June through October 1953 at Quonset Pt. and an AD-4Q was assigned to CVA-20 from October 1955 to April 1956. A small number of AD-4Ns were also flown from February through May 1954. Additionally, one S2F-2 was operated

from May 1959 to March 1960. The WF-2 began replacing the Skyraider in November 1959 and had completely replaced it in August 1961.

With the outbreak of the Korean War, VC-12 was called upon to supplement VC-11 aboard Pacific based carriers with three AEW/ASW detachments. These were: VC-12 Det 3 (AD-3Ws) aboard CV-32 from 6 September 1950 to 3 February 1951, LT L.B. Cornell commanding; Det 41 (AD-4Ws) aboard CVA-31 from 20 May 1952 to 8 January 1953, LCDR C.M. Blanchard commanding; and Det 44 (AD-4Ws) aboard CVA-39 from 26 April to 4 December 1953, LCDR D.Q. Joralmon commanding.

Training accidents were significant during the Skyraider's first ten

years of operations from NAS Quonst Pt. as outlined below. During night carrier qualifications off Quonset Pt. on 24 July 1951, LT John R. Wagner (AD-4W BuNo 124082) and ENS Thomas C. Reed (AD-4W BuNo 124105) collided in the downwind carrier landing pattern. Both pilots were killed and the aircraft were lost. Fifteen other accidents occurred during operations from Quonset through 1958. Only one occurred in 1952. On 19 December, LT Marion's left-hand main gear collapsed on landing

---

Below, VC-12 AD-4W BuNo 126857 at the Cleveland National Air Races in 1951. (Art Krieger via Jim Sullivan)

resulting in an underwing fire in AD-4 BuNo 124078. Another landing gear collapse occurred on 17 November 1953 resulting in minor damage to AD-4W BuNo 124766. Two other Skyraiders were also lost in 1953, both due to engine failures. The first was AD-4N, BuNo 125769, flown by ENS Robert A. Vankluyve. His engine failed during field carrier landing practice (FCLP) at NALF Charlestown on 2 June 1953. The aircraft crash landed in the woods after the engine exploded and the pilot escaped with minor injuries. The second failure occurred on 25 June 1953 off Narraganset Bay in AD-4W BuNo 124095. LCDR Micheal Baring RN and AN Joseph K. Kepple, Jr. were injured after ditching at sea. In 1954, three more gear failures occurred. On 1 April 1954, the landing gear failed on AD-3 BuNo 122901 during FCLP with ENS J.N. Babbitt at the controls. Six days later, the gear failed on AD-3W BuNo 122893 being flown by

ENS G. Godney. The damage was severe and the aircraft was salvaged. On 13 August 1954, LT W.B. Paradis had the tail gear of AD-3W, BuNo 122901, collapse during an emergency landing. On 20 July 1954, LTJG Litwin hit the water on take off and crashed into Narraganset Bay in AD-4W, BuNo 126865. Litwin

Above, VC-12 Det 11 pose aboard the USS Wright (CVL-49) from 11 January to 31 March 1951 for a cruise to the Med. LT D.L. Laird was Officer in Charge. (USN) Below, three VC-12 AD-3Ws overfly the USS Midway (CVA-41) over the Med on 16 June 1954. (National Archives)

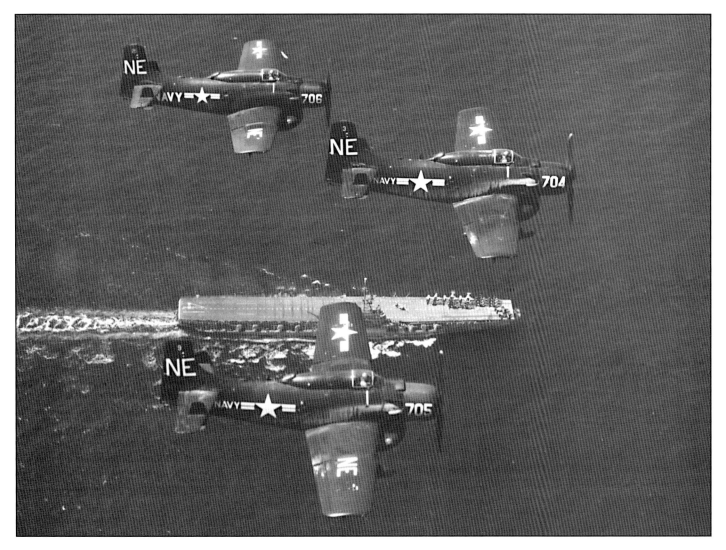

At top, VC-12 AD-5W BuNo 132772. (via Tailhook) Above, VC-12 AD-5W BuNo 133764 in flight. (Paul Minert collection) Below, VC-12 flightline with AD-5W BuNo 132789 in the foreground. Rudders have a thin white stripe. (Paul Minert collection)

escaped the aircraft without injury. LTJG Johnston blew a tire in AD-5W, BuNo 132789, on 1 August 1955 and ran off the runway damaging the prop and the right-hand wing. Two landing accidents were logged in 1956. LTJG

J.I. Becker and LTJG S.P. Swartz crash landed into a snow bank at Quonset Pt. on 1 April 1956 in AD-5W, BuNo 133773. Then, on 8 July 1956, LTJG J.W. Dempsey made a wheels-up landing in AD-5W, BuNo 133767. Both aircraft were returned to service. On 1 May 1958, LTJG William C. Cox declared a mayday eight miles west of Marthas Vineyard in an AD-5N. Pilot and aircraft were never found. The most unusual incident occurred in an AD-4W in 1954. LT Olmstead took off from Lawrence Airport with his wings folded! He proceeded to Quonset Pt. where he crash landed without injury off the end of the runway.

On 2 July 1956, VC-12 was

VC-12/VAW-12 Skyraider Assignments at NAS Quonset Pt. January 1950 to July 1961

VC-12:
Jan 49 AD-2 (3)
Feb 49 AD-2 (3)
Mar 49 AD-2 (3)
Apr 49 AD-2 (1), AD-1 (4)
May 49 AD-1 (4)
Jun 49 AD-1 (4)
Jul 49 AD-2 (5)
Aug 49 AD-3W (2), AD-3 (4), AD-2 (5)
Sep 49 AD-3W (11), AD-3 (4), AD-2 (5)
Oct 49 AD-3W (16), AD-3 (4), AD-2 (5)
Nov 49 AD-3W (20), AD-3 (4), AD-2 (5)
Dec 49 AD-3W (20), AD-3 (4), AD-5 (5)
Jan 50 AD-3W (20), AD-3 (4), AD-2 (5)
Feb 50 AD-3W (20), AD-3 (4), AD-2 (4)
Mar 50 AD-3W (15), AD-3 (4), AD-2 (4)
Apr 50 AD-3W (14), AD-3 (4), AD-2 (4)
May 50 AD-3W (15), AD-4 (4), AD-3 (8), AD-2 (4)
Jun 50 AD-4W (2), AD-3W (15), AD-4 (4), AD-3 (4), AD-2 (4)
Jul 50 AD-4W (5), AD-3W (15), AD-4 (2), AD-3 (3), AD-2 (3)
Aug 50 AD-4W (5), AD-3W (16), AD-4 (2), AD-3 (6)
Sep 50 AD-4W (5), AD-3W (3), AD-4 (2), AD-3 (2), AD-2 (3)
Oct 50 AD-4W (4), AD-3W (2), AD-4 (2), AD-3 (4), AD-2 (3)
Nov 50 AD-4W (8), AD-3W (8), AD-4 (2), AD-3 (8)
Dec 50 AD-4W (9), AD-3W (8), AD-3 (8), AD-2 (1)
Jan 51 AD-4W (10), AD-3W (4), AD-3 (8), AD-2 (1)
Feb 51 AD-4W (14), AD-3W (9), AD-3 (8), AD-2 (1)
Mar 51 AD-4W (12), AD-3W (7), AD-3 (8)
Apr 51 AD-4W (10), AD-3W (4), AD-3 (8)
May 51 AD-4W (12), AD-3W (4), AD-3 (8)
Jun 51 AD-4W (11), AD-3W (8), AD-3 (8)
Jul 51 AD-4W (10), AD-3W (3), AD-3 (8)
Aug 51 AD-4W (14), AD-3W (5), AD-3 (8)
Sep 51 AD-4W (8), AD-3W (5), AD-3 (8)
Oct 51 AD-4W (6), AD-3 (8)
Nov 51 AD-4W (10), AD-3W (6), AD-3 (8)
Dec 51 AD-4W (16), AD-3W (5), AD-3 (8)
Jan 52 AD-4W (8), AD-3W (2), AD-3 (7)
Feb 52 AD-4W (15), AD-3W (5), AD-3 (6)
Mar 52 AD-4W (11), AD-3W (5), AD-3 (6)
Apr 52 AD-4W (8), AD-3W (4), AD-3 (2), AD-1 (3)
May 52 AD-4W (4), AD-3W (3), AD-3 (1), AD-1 (4)
Jun 52 AD-4W (15), AD-3W (3), AD-3 (1), AD-1 (4)
Jul 52 AD-4W (18), AD-3W (4), AD-3 (1), AD-1 (4)
Aug 52 AD-4W (17), AD-3W (4), AD-3 (1), AD-1 (4)
Sep 52 AD-4W (21), AD-3W (4), AD-3 (3), AD-1 (4)
Oct 52 AD-4W (25), AD-3W (4), AD-3 (3)
Nov 52 AD-4W (18), AD-3W (4), AD-3 (2)
Dec 52 AD-4W (23), AD-3W (4), AD-3 (2)
Jan 53 AD-4W (24), AD-3W (4), AD-3 (2), AD-1 (1)
Feb 53 AD-4W (22), AD-3W (4), AD-3 (2), AD-1 (2)
Mar 53 AD-4W (27), AD-3W (4), AD-3 (2), AD-1 (1)
Apr 53 AD-4W (19), AD-3W (4), AD-3 (1)
May 53 AD-4W (18), AD-3W (2)
Jun 53 AD-4W (15), AD-3W (4), AD-3Q (1)
Jul 53 AD-4W (17), AD-3W (8), AD-3Q (1)
Aug 53 AD-4W (21), AD-3W (8), AD-3Q (1)
Sep 53 AD-4W (22), AD-3W (7), AD-3Q (1)
Oct 53 AD-4W (22), AD-3W (8), AD-3Q (1)
Nov 53 AD-4W (17), AD-3W (8)
Dec 53 AD-4W (25), AD-3W (7)
Jan 54 AD-4W (24), AD-3W (4), AD-3 (2), AD-1 (1)
Feb 54 AD-4W (15), AD-3W (9), AD-4N (4)
Mar 54 AD-4W (18), AD-3W (2), AD-4N (4)
Apr 54 AD-4W (19), AD-3W (6), AD-4N (3)
May 54 AD-4W (16), AD-3W (7), AD-4N (2)
Jun 54 AD-4W (19), AD-3W (8)
Jul 54 AD-4W (11), AD-3W (7)
Aug 54 AD-4W (19), AD-3W (8), AD-5 (4)
Sep 54 AD-4W (16), AD-3W (8), AD-5 (4)
Oct 54 AD-4W (13), AD-3W (3), AD-5 (5)
Nov 54 AD-4W (13), AD-3W (8), AD-5W (1), AD-5 (4)
Dec 54 AD-4W (12), AD-3W (6), AD-5W (6), AD-5 (4)

Jan 55 AD-4W (12), AD-3W (6), AD-5W (7), AD-5 (4)
Feb 55 AD-4W (12), AD-3W (6), AD-5W (7), AD-5 (4)
Mar 55 AD-4W (15), AD-3W (6), AD-5W (7), AD-5 (4)
Apr 55 AD-4W (17), AD-3W (4), AD-5W (14), AD-5 (4)
May 55 AD-4W (12), AD-5W (25)
Jun 55 AD-4W (4), AD-5W (29)
Jul 55 AD-4W (1), AD-5W (39)
Aug 55 AD-5W (35)
Sep 55 AD-5W (38)
Oct 55 AD-5W (30)
Nov 55 AD-5W (24), AD-4W (2)
Dec 55 AD-5W (23), AD-4W (3)
Jan 56 AD-5W (24)
Feb 56 AD-5W (26)
Mar 56 AD-5W (49)
Apr 56 AD-5W (39), AD-5 (2)
May 56 AD-5W (42), AD-5 (2)
Jun 56 AD-5W (45), AD-5 (2)

VAW-12:
Jul 56 AD-5W (36), AD-5 (2)
Aug 56 AD-5W (40), AD-5 (2)
Sep 56 AD-5W (39), AD-5 (2)
Oct 56 AD-5W (31), AD-5 (2)
Nov 56 AD-5W (30), AD-5 (2)
Dec 56 AD-5W (40), AD-5 (2)
Jan 57 AD-5W (37), AD-5 (2)
Feb 57 AD-5W (40), AD-5 (2)
Mar 57 AD-5W (40), AD-5 (2)
Apr 57 AD-5W (36), AD-5 (2)
May 57 AD-5W (44), AD-5 (2)
Jun 57 AD-5W (39), AD-5 (2)
Jul 57 AD-5W (43), AD-5 (2)
Aug 57 AD-5W (38), AD-5 (2)
Sep 57 AD-5W (36), AD-5 (1)
Oct 57 AD-5W (35), AD-5 (1)
Nov 57 AD-5W (37), AD-5 (1)
Dec 57 AD-5W (32), AD-5 (1)
Jan 58 AD-5W (39), AD-5 (1)
Feb 58 AD-5W (40), AD-5 (1)
Mar 58 AD-5W (40), AD-5 (1)
Apr 58 AD-5W (39), AD-5 (1)
May 58 AD-5W (37)
Jun 58 AD-5W (33)
Jul 58 AD-5W (23)
Aug 58 AD-5W (29)
Sep 58 AD-5W (22)
Oct 58 AD-5W (20)
Nov 58 AD-5W (21)
Dec 58 AD-5W (19)
Jan 59 AD-5W (18)
Feb 59 AD-5W (26)
Mar 59 AD-5W (26)
Apr 59 AD-5W (24)
May 59 AD-5W (22), S2F-2 (1)
Jun 59 AD-5W (19), S2F-2 (1)
Jul 59 AD-5W (20), S2F-2 (1)
Aug 59 AD-5W (12), S2F-2 (1)
Sep 59 AD-5W (21), S2F-2 (1)
Oct 59 AD-5W (28), AD-5 (2), S2F-2 (1)
Nov 59 AD-5W (25), AD-5 (2), S2F-2 (1), WF-2 (6)
Dec 59 AD-5W (23), AD-5 (2), S2F-2 (1), WF-2 (6)
Jan 60 AD-5W (16), AD-5 (2), S2F-2 (1), WF-2 (0)
Feb 60 AD-5W (23), AD-5 (2), S2F-2 (1), WF-2 (8)
Mar 60 AD-5W (19), AD-5 (2), WF-2 (10)
Apr 60 AD-5W (13), AD-5 (2), WF-2 (11)
May 60 AD-5W (13), AD-5 (2), WF-2 (11)
Jun 60 AD-5W (7), AD-5 (2), WF-2 (13)
Jul 60 AD-5W (10), AD-5 (2), WF-2 (12)
Aug 60 AD-5W (8), AD-5 (2), WF-2 (10)
Sep 60 AD-5W (14), AD-5 (2), WF-2 (13)
Oct 60 AD-5W (17), AD-5 (2), WF-2 (13)
Nov 60 AD-5W (10), AD-5 (1), WF-2 (13)
Dec 60 AD-5W (17), AD-5 (2), WF-2 (15)

VC-12   Det 32

CV-40   1951-53

Jan 61 AD-5W (18), AD-5 (2), WF-2 (7)
Feb 61 AD-5W (12), AD-5 (1), WF-2 (11)
Mar 61 AD-5W (8), WF-2 (17)
Apr 61 AD-5W (8), WF-2 (15)
May 61 AD-5W (10), WF-2 (13)
Jun 61 AD-5W (10), WF-2 (16)
Jul 61 AD-5W (10), WF-2 (15)
Aug 61 WF-2 (21)

VC-12/VAW-12 DETS 1950-1961

VC-12:

```
Sep 50 Det 3  Quonset  AD-3W (1)
       Det 4  CVL-28   AD-4W (1), AD-3W (1)
       Det 5  CVB-41   AD-3W (4)
       Det 6  CVB-43   AD-3W (4)
Oct 50 Det 3  CV-32    AD-3W (1)
       Det 4  CVL-28   AD-4W (1), AD-3W (1)
       Det 5  CVB-41   AD-3W (4)
       Det 6  CVB-43   AD-3W (4)
Nov 50 Det 3  CV-32    AD-3W (4)
       Det 6  CVB-43   AD-3W (4)
Dec 50 Det 3  CV-32    AD-3W (4)
       Det 6  CVB-43   AD-3W (3)
Jan 51 Det 3  CV-32    AD-3W (4)
       Det 6  CVB-43   AD-3W (3)
       Det 7  CVB-42   AD-4W (4)
       Det 11 CVL-49   AD-3W (3)
Feb 51 Det 3  CV-32    AD-3W (4)
       Det 7  CVB-42   AD-4W (4)
       Det 11 CVL-49   AD-3W (3)
Mar 51 Det 6  CVB-43   AD-4W (4)
       Det 7  CVB-42   AD-4W (4)
       Det 11 CVL-49   AD-3W (3)
Apr 51 Det 6  CVB-43   AD-4W (4)
       Det 7  CVB-42   AD-4W (4)
       Det 13 CVL-48   AD-3W (4)
May 51 Det 6  CVB-43   AD-4W (4)
       Det 13 CVL-48   AD-3W (4)
Jun 51 Det 6  CVB-43   AD-4W (4)
       Det 7  CVB-42   AD-4W (4)
       Det 32 CV-40    AD-4W (1)
Jul 51 Det 6  CVB-43   AD-4W (4)
       Det I  Norfolk  AD-4W (5)
       Det 14 CVE-112  AD-3W (4)
Aug 51 Det 5  CVB-41   AD-4W (2)
       Det 6  CVB-43   AD-4W (4)
Sep 51 Det 3  CV-32    AD-4W (4)
       Det 6  CVB-43   AD-4W (4)
       Det 8  CV-34    AD-4W (4)
Oct 51 Det 3  CV-32    AD-4W (4)
       Det 5  CVB-41   AD-4W (5), AD-3W (3)
       Det 7  CVB-42   AD-4W (4)
       Det 12 CVE-120  AD-4W (1), AD-3W (2)
       Det 13 CVL-48   AD-4W (1)
       Det 32 CV-40    AD-4W (4)
Nov 51 Det 3  CV-32    AD-4W (4)
       Det 5  CVB-41   AD-4W (5), AD-3W (3)
       Det 7  CVB-42   AD-4W (4)
       Det 13 CVL-48   AD-4W (1)
       Det 32 CV-40    AD-4W (4)
Dec 51 Det 3  CV-32    AD-4W (4)
       Det 5  CVB-41   AD-4W (5), AD-3W (3)
       Det 7  CVB-42   AD-4W (4)
       Det 13 CVL-48   AD-4W (1)
       Det 32 CV-40    AD-4W (4)
Jan 52 Det 3  CV-32    AD-4W (3), AD-3W (4)
       Det 5  CVB-41   AD-4W (4)
       Det 7  CVB-42   AD-4W (4)
       Det 32 CV-40    AD-4W (4)
Feb 52 Det 5  CVB-41   AD-4W (4)
       Det 32 CV-40    AD-4W (4)
Mar 52 Det 5  CVB-41   AD-4W (4)
       Det 6  CVB-43   AD-4W (3)
       Det 32 CV-40    AD-4W (4)
Apr 52 Det 5  CVB-41   AD-4W (4)
       Det 6  CVB-43   AD-4W (3)
       Det 32 CV-40    AD-4W (4)
       Det 38 Quonset  AD-4W (3)
May 52 Det 3  CV-32    AD-4W (8)
       Det 6  CVB-43   AD-4W (3)
       Det 32 CV-40    AD-4W (4)
       Det 38 CV-18    AD-4W (3)
       Det 41 CV-31    AD-4W (3)
Jun 52 Det 6  CVB-43   AD-4W (3)
       Det 7  CVB-42   AD-4W (3)
       Det 38 CV-18    AD-4W (2)
       Det 41 CV-31    AD-4W (3)
Jul 52 Det 6  CVB-43   AD-4W (3)
       Det 7  CVB-42   AD-4W (3)
       Det 38 CV-18    AD-4W (2)
       Det 41 CV-31    AD-4W (3)
Aug 52 Det 3  CV-32    AD-4W (3)
       Det 5  CVB-41   AD-4W (3)
       Det 6  CVB-43   AD-4W (3)
       Det 7  CVB-42   AD-4W (2)
       Det 38 CV-18    AD-4W (3)
       Det 41 CV-31    AD-4W (3)
Sep 52 Det 3  CV-32    AD-4W (3)
       Det 5  CVB-41   AD-4W (3)
       Det 6  CVB-43   AD-4W (3)
       Det 7  CVB-42   AD-4W (2)
       Det 38 CV-18    AD-4W (2)
       Det 41 CV-31    AD-4W (3)
Oct 52 Det 3  CVA-32   AD-4W (3)
       Det 5  CVA-41   AD-4W (3)
       Det 7  CVA-42   AD-4W (2)
       Det 41 CVA-31   AD-4W (3)
Nov 52 Det 3  CVA-32   AD-4W (3)
       Det 5  CVA-41   AD-4W (3)
       Det 7  CVA-42   AD-4W (2)
       Det 32 CVA-40   AD-4W (3)
       Det 41 CVA-31   AD-4W (3)
Dec 52 Det 3  CVA-32   AD-4W (3)
       Det 5  CVA-41   AD-4W (3)
       Det 7  CVA-42   AD-4W (2)
       Det 32 CVA-40   AD-4W (3)
       Det 41 CVA-31   AD-4W (3)
Jan 53 Det 3  CVA-32   AD-4W (2)
       Det 5  CVA-41   AD-4W (4)
       Det 32 CVA-40   AD-4W (4)
       Det 41 CVA-31   AD-4W (3)
Feb 53 Det 5  CVA-41   AD-4W (4)
       Det 7  CVA-42   AD-4W (6)
       Det 32 CVA-40   AD-4W (3)
Feb 53 Det 5  CVA-41   AD-4W (4)
       Det 7  CVA-42   AD-4W (6)
       Det 32 CVA-40   AD-4W (3)
Mar 53 Det 5  CVA-41   AD-4W (2)
       Det 7  CVA-42   AD-4W (1)
       Det 32 CVA-40   AD-4W (3)
Apr 53 Det 5  CVA-41   AD-4W (2)
       Det 6  CVA-43   AD-4W (3)
       Det 7  CVA-42   AD-4W (1)
       Det 32 CVA-40   AD-4W (3)
       Det 44 CVA-39   AD-4W (3)
May 53 Det 6  CVA-43   AD-4W (3)
       Det 7  CVA-42   AD-4W (3)
       Det 9  Quonset  AD-3Q (1)
       Det 32 CVA-40   AD-4W (3)
       Det 39 CVA-36   AD-4W (1)
       Det 44 CVA-39   AD-4W (3)
Jun 53 Det 6  CVA-43   AD-4W (3)
       Det 7  CVA-42   AD-4W (3)
       Det 9  Quonset  AD-3Q (1)
       Det 11 CVL-49   AD-4W (2), AD-3W (4)
       Det 32 CVA-40   AD-4W (1)
       Det 39 CVA-36   AD-4W (1)
       Det 44 CVA-39   AD-4W (3)
Jul 53 Det 6  CVA-43   AD-4W (3)
       Det 7  CVA-42   AD-4W (3)
       Det 9  Quonset  AD-3Q (1)
       Det 39 CVA-36   AD-4W (1)
       Det 41 Quonset  AD-4W (2)
       Det 44 CVA-39   AD-4W (3)
Aug 53 Det 6  CVA-43   AD-4W (2)
       Det 7  CVA-42   AD-4W (3)
       Det 9  Quonset  AD-3Q (1)
       Det 41 Quonset  AD-4W (3)
       Det 44 CVA-39   AD-4W (3)
```

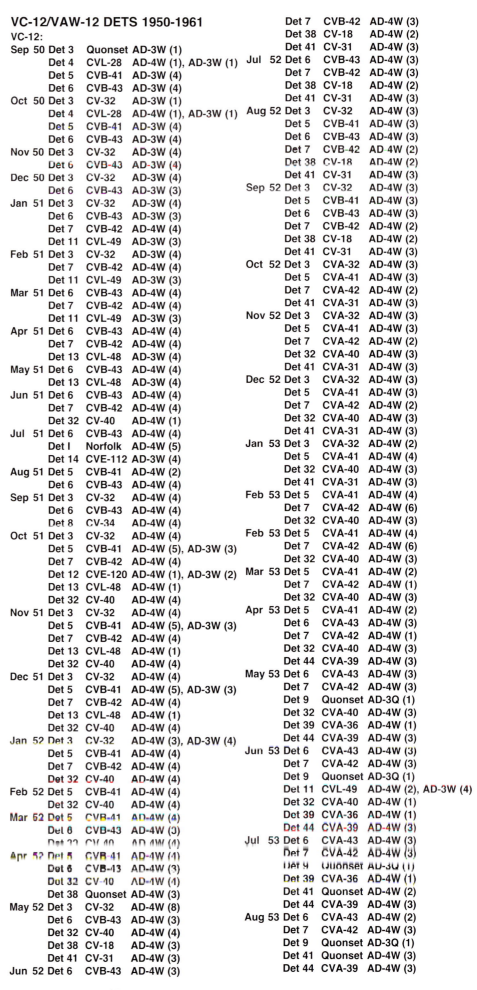

CVA-39   1956

Above, VAW-12 AD-5W BuNo 135219 in flight from Forrestal on 11 October 1958. (USN via Barry Miller) At right, VAW-12 AD-5W BuNo 132758 with the squadron's new "GE" tail code assigned on 1 July 1957. (Ginter collection) Below, VAW-12 AD-5W BuNo 132789 trapping. (Paul Minert collection)

redesignated Airborne Early Warning Squadron Twelve (VAW-12). The squadron's "NE" tail code was changed to "GE" on 1 July 1957.

Date	Det	Ship	Aircraft
	Det 53	CVA-45	AD-4W (2)
Sep 53	Det 6	CVA-43	AD-4W (2)
	Det 7	CVA-42	AD-4W (3)
	Det 9	Quonset	AD-3Q (1)
	Det 38	CVA-18	AD-4W (3)
	Det 41	Quonset	AD-4W (3)
	Det 44	CVA-39	AD-4W (3)
Oct 53	Det 7	CVA-42	AD-4W (3)
	Det 32	CVA-40	AD-4W (3)
	Det 38	CVA-18	AD-4W (3)
	Det 41	Quonset	AD-4W (3)
	Det 44	CVA-39	AD-4W (3)
Nov 53	Det 5	CVA-41	AD-4W (3)
	Det 7	CVA-42	AD-4W (3)
	Det 32	CVA-40	AD-4W (3)
	Det 38	CVA-18	AD-4W (3)
	Det 41	Quonset	AD-4W (3)
	Det 44	CVA-39	AD-4W (3)
Dec 53	Det 5	CVA-41	AD-4W (3)
	Det 32	CVA-40	AD-4W (3)
	Det 38	CVA-18	AD-4W (3)
	Det 41	Quonset	AD-4W (3)
Jan 54	Det 3	CVS-32	AD-4W (2)
	Det 5	CVA-41	AD-4W (3)
	Det 32	CVA-40	AD-4W (3)
	Det 38	CVA-18	AD-4W (3)
Feb 54	Det 5	CVA-41	AD-4W (3)
	Det 38	CVA-40	AD-4W (3)
	Det 54	CVA-15	AD-4W (3)
	Det 57	CVA-12	AD-4W (3)
Mar 54	Det 35	CVA-41	AD-4W (3)
	Det 38	CVA-40	AD-4W (3)
	Det 48	CVL-49	AD-4W (2), AD-3W (7)
Apr 54	Det 35	CVA-41	AD-4W (3)
	Det 38	CVA-40	AD-4W (3)
	Det 40	CVA-18	AD-4W (1)
May 54	Det 30	CVA-20	AD-4W (3)
	Det 31	CVA-43	AD-4W (1)
	Det 35	CVA-41	AD-4W (3)
	Det 38	CVA-40	AD-4W (3)
Jun 54	Det 31	CVA-43	AD-4W (1)
	Det 35	CVA-41	AD-4W (3)
	Det 38	CVA-40	AD-4W (3)
Jul 54	Det 31	CVA-43	AD-4W (3)
	Det 35	CVA-41	AD-4W (3)
	Det 38	CVA-40	AD-4W (3)
	Det 52	CVS-45	AD-4W (5), AD-3W (1)
Aug 54	Det 31	CVA-43	AD-4W (3)
	Det 34	CVA-39	AD-4W (2)
Sep 54	Det 31	CVA-43	AD-4W (3)
	Det 34	CVA-39	AD-4W (2)
Oct 54	Det 31	CVA-43	AD-4W (3)
	Det 34	CVA-39	AD-4W (3)
	Det 35	CVA-41	AD-4W (3)
Nov 54	Det 31	CVA-43	AD-4W (3)
	Det 34	CVA-39	AD-4W (3)
	Det 35	CVA-41	AD-4W (3)
Dec 54	Det 34	CVA-39	AD-4W (3)
	Det 35	CVA-41	AD-4W (3)
Jan 55	Det 34	CVA-39	AD-4W (3)
	Det 35	CVA-41	AD-4W (3)
	Det 36	CVA-15	AD-4W (3)
Feb 55	Det 34	CVA-39	AD-4W (3)
	Det 35	CVA-41	AD-4W (3)
	Det 36	CVA-15	AD-4W (3)
Mar 55	Det 31	CVA-43	AD-4W (3)
	Det 34	CVA-39	AD-4W (3)
	Det 35	CVA-41	AD-4W (3)
	Det 36	CVA-15	AD-4W (3)
	Det 39	CVA-14	AD-5W (3)
Apr 55	Det 31	CVA-43	AD-4W (3)
	Det 32	Miramar	AD-4W (3)
	Det 35	CVA-41	AD-4W (3)
	Det 36	CVA-15	AD-4W (3)
	Det 39	CVA-14	AD-5W (3)
May 55	Det 31	CVA-43	AD-4W (3)
	Det 32	CVA-12	AD-4W (3)
	Det 33	CVA-11	AD-4W (4)
	Det 35	CVA-41	AD-4W (3)
	Det 36	CVA-15	AD-4W (4)
Jun 55	Det 30	CVA-20	AD-5W (3)
	Det 31	CVA-43	AD-4W (3)
	Det 32	CVA-12	AD-4W (3)
	Det 33	CVA-11	AD-4W (3)
	Det 35	CVA-41	AD-4W (4)
Jul 55	Det 31	CVA-43	AD-4W (3)
	Det 32	CVA-12	AD-4W (3)
	Det 33	CVA-11	AD-4W (3)
Aug 55	Det 31	CVA-43	AD-4W (3)
	Det 32	CVA-12	AD-4W (3)
	Det 33	CVA-11	AD-4W (2)
	Det 34	CVA-39	AD-5W (4)
Sep 55	Det 31	CVA-43	AD-4W (3)
	Det 32	CVA-12	AD-4W (3)
	Det 33	CVA-11	AD-4W (3)
	Det 34	CVA-39	AD-5W (3)
Oct 55	Det 30	CVA-20	AD-5W (3), AD-4Q (1)
	Det 32	CVA-12	AD-4W (3)
	Det 33	CVA-11	AD-4W (2)
	Det 34	CVA-39	AD-5W (3)
	Det 39	CVA-14	AD-5W (5)
Oct 55	Det 30	CVA-20	AD-5W (3), AD-4Q (1)
	Det 32	CVA-12	AD-4W (3)
	Det 33	CVA-11	AD-4W (2)
	Det 34	CVA-39	AD-5W (3)
	Det 39	CVA-14	AD-5W (5)
Nov 55	Det 30	CVA-20	AD-5W (1), AD-4Q (1)
	Det 32	CVA-12	AD-4W (3)
	Det 34	CVA-39	AD-5W (3)
	Det 39	CVA-14	AD-5W (5)
Dec 55	Det 30	CVA-20	AD-5W (3), AD-4Q (1)
	Det 34	CVA-39	AD-5W (3)
	Det 39	CVA-14	AD-5W (5)
Jan 56	Det 30	CVA-20	AD-5W (3), AD-4Q (1)
	Det 34	CVA-39	AD-5W (3)
	Det 39	CVA-14	AD-5W (5)
Feb 56	Det 30	CVA-20	AD-5W (3), AD-4Q (1)
	Det 34	CVA-39	AD-5W (3)
	Det 39	CVA-14	AD-5W (5)
Mar 56	Det 30	CVA-20	AD-5W (3), AD-4Q (1)
	Det 31	CVA-43	AD-5W (2)
	Det 36	CVA-15	AD-5W (4)
	Det 39	CVA-14	AD-5W (5)
Apr 56	Det 39	CVA-14	AD-5W (4)
May 56	Det 39	CVA-14	AD-5W (5)
Jun 56	Det 39	CVA-14	AD-5W (4)

VAW-12:

Date	Det	Ship	Aircraft
Jul 56	Det 31	CVA-43	AD-5W (4)
	Det 36	CVA-15	AD-5W (4)
	Det 39	CVA-14	AD-5W (4)
Aug 56	Det 31	CVA-43	AD-5W (4)
	Det 36	CVA-15	AD-5W (4)
	Det 39	CVA-14	AD-5W (4)
Sep 56	Det 30	CVA-20	AD-5W (4)
	Det 31	CVA-43	AD-5W (3)
	Det 36	CVA-15	AD-5W (4)
Oct 56	Det 30	CVA-20	AD-5W (4)
	Det 31	CVA-43	AD-5W (3)
	Det 36	CVA-15	AD-5W (4)
	Det 42	CVA-59	AD-5W (6)
Nov 56	Det 30	CVA-20	AD-5W (1)
	Det 31	CVA-43	AD-5W (4)
	Det 34	CVA-39	AD-5W (4)
	Det 36	CVA-15	AD-5W (4)
	Det 42	CVA-59	AD-5W (4)
Dec 56	Det 30	CVA-20	AD-5W (4)
	Det 31	CVA-43	AD-5W (4)
	Det 34	CVA-39	AD-5W (4)
	Det 36	CVA-15	AD-5W (4)
	Det 42	CVA-59	AD-5W (4)
Jan 57	Det 30	CVA-20	AD-5W (4)
	Det 31	CVA-43	AD-5W (4)
	Det 34	CVA-39	AD-5W (4)
	Det 36	CVA-15	AD-5W (4)
	Det 42	CVA-59	AD-5W (4)
Feb 57	Det 30	CVA-20	AD-5W (4)
	Det 31	CVA-43	AD-5W (4)
	Det 34	CVA-39	AD-5W (4)
	Det 42	CVA-59	AD-5W (4)
Mar 57	Det 30	CVA-20	AD-5W (4)
	Det 34	CVA-39	AD-5W (4)
	Det 42	CVA-59	AD-5W (4)
May 57	Det 34	CVA-39	AD-5W (4)
	Det 36	CVA-15	AD-5W (4)
	Det 42	CVA-59	AD-5W (4)
Jun 57	Det 34	CVA-39	AD-5W (4)
	Det 36	CVA-15	AD-5W (4)
	Det 42	CVA-59	AD-5W (4)
Jul 57	Det 33	CVA-11	AD-5W (4)
	Det 34	CVA-39	AD-5W (3)
	Det 36	CVA-15	AD-5W (4)
	Det 42	CVA-59	AD-5W (4)
Aug 57	Det 33	CVA-11	AD-5W (4)
	Det 36	CVA-15	AD-5W (4)
	Det 42	CVA-59	AD-5W (5)
Sep 57	Det 33	CVA-11	AD-5W (4)
	Det 36	CVA-15	AD-5W (5)
Oct 57	Det 33	CVA-11	AD-5W (4)
	Det 36	CVA-15	AD-5W (4)
Nov 57	Det 36	CVA-15	AD-5W (4)
	Det 45	CVA-9	AD-5W (4)
Dec 57	Det 36	CVA-15	AD-5W (4)
	Det 45	CVA-9	AD-5W (2)
Jan 58	Det 36	CVA-15	AD-5W (4)
	Det 45	CVA-9	AD-5W (4)
Feb 58	Det 45	CVA-9	AD-5W (4)
Mar 58	Det 45	CVA-9	AD-5W (4)
Apr 58	Det 42	CVA-59	AD-5W (4)
	Det 45	CVA-9	AD-5W (4)
May 58	Det 33	CVA-11	AD-5W (4)
	Det 45	CVA-9	AD-5W (4)
	Det 48	CVL-49	AD-5W (4)
Jun 58	Det 33	CVA-11	AD-5W (4)
	Det 36	CVA-15	AD-5W (4)
	Det 45	CVA-9	AD-5W (4)
	Det 48	CVL-49	AD-5W (4)
Jul 58	Det 33	CVA-11	AD-5W (4)
	Det 36	CVA-15	AD-5W (4)
	Det 42	CVA-59	AD-5W (4)
	Det 45	CVA-9	AD-5W (4)
	Det 48	CVS-18	AD-5W (4)
	Det 52	CVS-45	AD-5W (6)
Aug 58	Det 36	CVA-15	AD-5W (4)
	Det 42	CVA-59	AD-5W (4)
	Det 45	CVA-9	AD-5W (5)
	Det 48	CVS-18	AD-5W (4)
	Det 52	CVS-45	AD-5W (5)
Sep 58	Det 34	CVS-39	AD-5W (4)
	Det 36	CVA-15	AD-5W (8)
	Det 42	CVA-59	AD-5W (4)
	Det 45	CVA-9	AD-5W (1)
	Det 48	CVS-18	AD-5W (4)
	Det 52	CVS-45	AD-5W (5)
Oct 58	Det 33	CVA-11	AD-5W (3)
	Det 34	CVS-39	AD-5W (4)
	Det 36	CVA-15	AD-5W (4)
	Det 42	CVA-59	AD-5W (4)
	Det 45	CVA-9	AD-5W (5)
	Det 48	CVS-18	AD-5W (4)
	Det 51	CVS-32	AD-5W (1)
	Det 52	CVS-45	AD-5W (5)
Nov 58	Det 33	CVA-11	AD-5W (4)
	Det 34	CVS-39	AD-5W (4)
	Det 36	CVA-15	AD-5W (4)
	Det 38	Quonset	AD-5W (4)

Det 42 CVA-59 AD-5W (4)
Det 45 CVA-9 AD-5W (1)
Det 48 CVS-18 AD-5W (4)
Det 51 CVS-32 AD-5W (1)
Det 52 CVS-45 AD-5W (5)
Dec 58 Det 33 CVA-11 AD-5W (5)
Det 34 CVS-39 AD-5W (4)
Det 36 CVA-15 AD-5W (4)
Det 38 Quonset AD-5W (3)
Det 42 CVA-59 AD-5W (4)
Det 45 CVA-9 AD-5W (1)
Det 48 CVS-18 AD-5W (4)
Det 51 CVS-32 AD-5W (1)
Det 52 CVS-45 AD-5W (4)
Jan 59 Det 33 CVA-11 AD-5W (4)
Det 34 CVS-39 AD-5W (4)
Det 36 CVA-15 AD-5W (4)
Det 38 Quonset AD-5W (4)
Det 42 CVA-59 AD-5W (4)
Det 45 CVA-9 AD-5W (1)
Det 48 CVS-18 AD-5W (4)
Det 51 CVS-32 AD-5W (2)
Det 52 CVS-45 AD-5W (4)
Feb 59 Det 33 CVA-11 AD-5W (4)
Det 34 CVS-39 AD-5W (3)
Det 36 CVA-15 AD-5W (3)
Det 38 Quonset AD-5W (4)
Det 42 CVA-59 AD-5W (4)
Det 45 CVA-9 AD-5W (1)
Det 48 CVS-18 AD-5W (3)
Det 52 CVS-45 AD-5W (5)
Mar 59 Det 33 CVA-11 AD-5W (4)
Det 34 CVS-39 AD-5W (3)
Det 36 CVS-15 AD-5W (3)
Det 38 CVS-40 AD-5W (3)
Det 45 CVA-9 AD-5W (1)
Det 48 CVS-18 AD-5W (3)
Det 52 CVS-45 AD-5W (3)
Apr 59 Det 33 CVA-11 AD-5W (4)
Det 34 CVS-39 AD-5W (3)
Det 36 CVS-15 AD-5W (3)
Det 38 CVS-40 AD-5W (5)
Det 45 CVA-9 AD-5W (3)
Det 48 CVS-18 AD-5W (3)
Det 52 CVS-45 AD-5W (3)
May 59 Det 33 CVA-11 AD-5W (4)
Det 34 CVS-39 AD-5W (4)
Det 36 CVS-15 AD-5W (3)
Det 38 CVS-40 AD-5W (3)
Det 45 CVA-9 AD-5W (2)
Det 48 CVS-18 AD-5W (3)
Det 52 CVS-45 AD-5W (4)
Jun 59 Det 33 CVA-11 AD-5W (4)
Det 34 CVS-39 AD-5W (4)
Det 36 CVS-15 AD-5W (4)
Det 38 CVS-40 AD-5W (4)
Det 45 CVA-9 AD-5W (3)
Det 48 CVS-18 AD-5W (4)
Det 52 CVS-45 AD-5W (4)
Jul 59 Det 33 CVA-11 AD-5W (4)
Det 34 CVS-39 AD-5W (4)
Det 36 CVS-15 AD-5W (4)
Det 38 CVS-40 AD-5W (4)
Det 45 CVA-9 AD-5W (2)
Det 48 CVS-18 AD-5W (4)
Det 52 CVS-45 AD-5W (2)
Aug 59 Det 33 CVA-11 AD-5W (4), AD-5Q (1)
Det 34 CVS-39 AD-5W (5)
Det 36 CVS-15 AD-5W (4)
Det 38 CVS-40 AD-5W (4)
Det 41 CVA-62 AD-5W (1)
Det 42 CVA-59 AD-5W (1)
Det 43 CVA-60 AD-5W (1)
Det 45 CVA-9 AD-5W (3)
Det 48 CVS-18 AD-5W (3)

Det 52 CVS-45 AD-5W (4)
Sep 59 Det 34 CVS-39 AD-5W (4)
Det 36 CVS-15 AD-5W (4)
Det 38 CVS-40 AD-5W (4)
Det 42 CVA-59 AD-5W (1)
Det 43 CVA-60 AD-5W (1)
Det 45 CVA-9 AD-5W (3)
Det 48 CVS-18 AD-5W (4)
Oct 59 Det 36 CVS-15 AD-5W (4)
Det 42 CVA-59 AD-5W (1)
Det 43 CVA-60 AD-5W (1)
Det 45 CVA-9 AD-5W (3)
Det 48 CVS-18 AD-5W (3)
Nov 59 Det 38 CVS-40 AD-5W (4)
Det 42 CVA-59 AD-5W (1)
Det 43 CVA-60 AD-5W (1)
Det 45 CVA-9 AD-5W (3)
Det 52 CVS-45 AD-5W (4)
Dec 59 Det 42 CVA-59 AD-5W (1)
Det 43 CVA-60 AD-5W (1)
Det 45 CVA-9 AD-5W (3)
Det 48 CVS-18 AD-5W (4)
Det 52 CVS-45 AD-5W (4)
Jan 60 Det 36 CVS-15 AD-5W (1)
Det 42 CVA-59 AD-5W (4)
Det 43 CVA-60 AD-5W (3)
Det 45 CVA-9 AD-5W (3)
Det 48 CVS-18 AD-5W (4)
Det 52 CVS-45 AD-5W (4)
Feb 60 Det 34 CVS-39 AD-5W (5)
Det 36 CVS-15 AD-5W (1)
Det 42 CVA-59 AD-5W (4)
Det 45 CVA-9 AD-5W (1)
Det 48 CVS-18 AD-5W (1)
Det 52 CVS-45 AD-5W (2)
Mar 60 Det 34 CVS-39 AD-5W (5)
Det 36 CVS-15 AD-5W (4)
Det 42 CVA-59 AD-5W (4)
Det 52 CVS-45 AD-5W (5)
Apr 60 Det 34 CVS-39 AD-5W (5)
Det 36 CVS-15 AD-5W (5)
Det 42 CVA-59 AD-5W (4)
Det 52 CVS-45 AD-5W (4)
May 60 Det 36 CVS-15 AD-5W (4)
Det 42 CVA-59 AD-5W (4)
Det 45 CVS-9 AD-5W (4)
Det 52 CVS-45 AD-5W (5)
Jun 60 Det 36 CVS-15 AD-5W (4)
Det 42 CVA-59 AD-5W (4)
Det 45 CVS-9 AD-5W (4)
Det 48 CVS-18 AD-5W (4)
Det 52 CVS-45 AD-5W (5)
Jul 60 Det 34 CVS-39 AD-5W (2)
Det 36 CVS-15 AD-5W (4)
Det 41 CVA-62 WF-2 (5)
Det 45 CVS-9 AD-5W (4)
Det 48 CVS-18 AD-5W (4)
Det 52 CVS-45 AD-5W (4)
Aug 60 Det 34 CVS-39 AD-5W (5)
Det 36 CVS-15 AD-5W (4)
Det 41 CVA-62 WF-2 (5)
Det 43 CVA-60 WF-2 (5)
Det 45 CVS-9 AD-5W (3)
Det 48 CVS-18 AD-5W (4)
Det 52 CVS-45 AD-5W (4)
Sep 60 Det 34 CVS-39 AD-5W (1)
Det 41 CVA-62 WF-2 (5)
Det 43 CVA-60 WF-2 (5)
Det 45 CVS-9 AD-5W (3)
Det 48 CVS-18 AD-5W (4)
Det 52 CVS-45 AD-5W (4)
Oct 60 Det 34 CVS-39 AD-5W (1)
Det 41 CVA-62 WF-2 (5)
Det 43 CVA-60 WF-2 (5)
Det 45 CVS-9 AD-5W (4)

Det 48 CVS-18 AD-5W (4)
Det 52 CVS-45 AD-5W (5)
Nov 60 Det 34 CVS-39 AD-5W (4)
Det 41 CVA-62 WF-2 (5)
Det 43 CVA-60 WF-2 (5)
Det 45 CVS-9 AD-5W (4)
Det 48 CVS-18 AD-5W (4)
Det 52 CVS-45 AD-5W (5)
Dec 60 Det 34 CVS-39 AD-5W (4)
Det 41 CVA-62 WF-2 (5)
Det 43 CVA-60 WF-2 (5)
Det 52 CVS-45 AD-5W (4)
Jan 61 Det 34 CVS-39 AD-5W (3)
Det 37 CVA-42 WF-2 (3)
Det 41 CVA-62 WF-2 (5)
Det 42 CVA-59 WF-2 (6)
Det 43 CVA-60 WF-2 (5)
Det 52 CVS-45 AD-5W (4)
Feb 61 Det 34 CVS-39 AD-5W (3)
Det 37 CVA-42 WF-2 (3)
Det 41 CVA-62 WF-2 (5)
Det 42 CVA-59 WF-2 (5)
Det 48 CVS-18 AD-5W (4)
Det 52 CVS-45 AD-5W (3)
Mar 61 Det 34 CVS-39 AD-5W (3)
Det 37 CVA-42 WF-2 (3)
Det 42 CVA-59 WF-2 (5)
Det 45 CVS-9 AD-5W (4)
Det 52 CVS-45 AD-5W (1)
Apr 61 Det 34 CVS-39 AD-5W (5)
Det 36 CVS-15 WF-2 (5)
Det 37 CVA-42 WF-2 (2)
Det 42 CVA-59 WF-2 (5)
Det 45 CVS-9 AD-5W (4)
Det 48 CVS-18 AD-5W (4)
Det 52 CVS-45 AD-5W (1)
May 61 Det 34 CVS-39 AD-5W (4)
Det 37 CVA-42 WF-2 (3)
Det 41 CVA-62 WF-2 (4)
Det 42 CVA-59 WF-2 (5)
Det 48 CVS-18 AD-5W (4)
Jun 61 Det 34 CVS-39 AD-5W (4)
Det 37 CVA-42 WF-2 (3)
Det 41 CVA-62 WF-2 (5)
Det 42 CVA-59 WF-2 (5)
Det 48 CVS-18 AD-5W (4)
Jul 61 Det 37 CVA-42 WF-2 (3)
Det 41 CVA-62 WF-2 (4)
Det 42 CVA-59 WF-2 (5)
Det 48 CVS-18 AD-5W (4)
Aug 61 Det 36 CVS-15 WF-2 (4)
Det 41 CVA-62 WF-2 (4)

Then as previously stated, the WF-2 had completely replaced the Skyraider in August 1961. In 1964, E-2A Hawkeyes began replacing the WF-2/EA-1B and on 1 April 1967, VAW-12 became Carrier Airborne Early Warning Wing Twelve and its assets formed VAW-121 with EA-1Bs and VAW-122 and VAW-123 with E-2As.

VAW-12

Above, VAW-12 AD-5W BuNo 135180 was assigned to the USS Essex (CVS-9). Note Essex on top of tail written in script. Bat on tail was maroon and belly radome has been removed. (Ginter collection) At left, three VAW-12 Det 45 AD-5Ws from the USS Essex (CVA-9) in September 1958. AD-5W BuNo 132765 (AP/701) was crewed by LTJG E.A. Cooper, LTJG R.D. Sundbye, and AN M.D. Timothy; 135191 (AP-704) crewed by LTJG J.E. Shay and AD3 D.L. Cyr; 132743 (AP-702) crewed by LTJG S.L. Zwick, LTJG C.P. Mooney, and AT2 G.E. Rojan. (USN) Bottom, VAW-12 AD-5W BuNo 139580 at Litchfield Park, AZ, in May 1960. Stylized bat on the tail was maroon outlined in yellow. (William Swisher)

AIRBORNE EARLY WARNING SQUADRON THIRTEEN, VAW-13 "ZAPPERS" ELECTRONIC WARFARE SQUADRON ONE - HUNDRED - THIRTY, VAQ-130

Airborne Early Warning Squadron Thirteen (VAW-13) was established on 1 September 1959 at NAS Agana, Guam, with three AD-5Ws. By month end three more AD-5Ws and crews were received from VAW-11 and two three-plane Dets were formed. These were Det A and Det D. By the end of October, Det A was aboard the USS Kearsarge (CVS-33) and Det D was at Cubi Pt. Acting CO was LCDR Henry Kriszamer who was replaced in February 1960 by CDR E.K. Hitchcock Jr.

VAW-13 received its first two AD-5Qs in December 1959 and its first four WF-2s in August 1960. By the end of February 1961, all the AD-5Ws

At right, VAW-13 Det M AD-5W BuNo 132767 landing aboard the USS Ranger (CVA-61) in 1960. (USN) Below, VAW-13 AD-5Q BuNo 132580 (VR/704) landing at NAS Atsugi, Japan, on 24 July 1961 was assigned to CVG-5. Aircraft was assigned to the USS Ticonderoga (CVA-14) as part of CVG-5. (Toyokazu Matsuzaki)

had been replaced with Grumman WF-2s and in June, the WF-2s were transferred out due to a change in mission. VAW-13 became AirPacs electronic countermeasures squadron and transferred its WF-2s to VAW-11. It then took custody of all Pacific based AD-5Q "Queer Spads"/"Electric Spads". In July a TF-1Q was acquired. A second TF-1Q was received in August and the squadron transferred to NAS Alameda, CA, that same month.

In 1962, in an effort to stop night-time supply drops over the central highlands in South Vietnam, the Air Force and the Navy were tasked with the interception of these aircraft. The aircraft were thought to be the AN-2 Colt, a single radial-engined rather large biplane. The Air Force answer was the F-102 Delta Dart, the Navy's answer was VAW-13's AD-5Q/EA-1F. On the surface, the Skyraider was a much wiser choice as it could match the AN-2's flight characteristics and 20mm guns should work better than

the F-102's Falcons or 2.75" rockets. Since no aircraft were ever intercepted, we will never know which aircraft was better for the mission. In any event, on 7 May 1962, VAW-13 was ordered to establish land-based oper-

VAW-13/VAQ-130 ASSIGNMENTS 1959-1968

VAW-13:

Date	Det	Location	Aircraft
Sep 59	Det A	Cubi	AD-5W (3)
	Det D	Cubi	AD-5W (3)
Oct 59	Det A	CVS-33	AD-5W (3)
	Det D	Cubi	AD-5W (3)
Nov 59	Det	Agana	AD-5W (5)
	Det A	CVS-33	AD-5W (3)
Dec 59	Det	Agana	AD-5W (5), AD-5Q (2)
	Det A	CVS-33	AD-5W (3)
Jan 60	Det	Agana	AD-5W (7), AD-5Q (2)
	Det A	CVS-33	AD-5W (3)
Feb 60	Det	Agana	AD-5W (5), AD-5Q (4)
	Det A	Kisarazu	AD-5W (3)
Mar 60	Det	Agana	AD-5W (8), AD-5Q (4)
	Det M	CVA-61	AD-5W (3), AD-5Q (2)
Apr 60	Det	Agana	AD-5W (8), AD-5Q (4)
	Det M	CVA-61	AD-5W (3), AD-5Q (2)
May 60	Det	Agana	AD-5W (8), AD-5Q (4)
	Det M	CVA-61	AD-5W (3), AD-5Q (2)
Jun 60	Det	Agana	AD-5W (8), AD-5Q (6)
	Det M	CVA-61	AD-5W (3), AD-5Q (2)
Jul 60	Det	Agana	AD-5W (8), AD-5Q (4)
	Det F	CVA-34	AD-5Q (2)
	Det M	CVA-61	AD-5W (3), AD-5Q (2)
	Det N	CVS-12	AD-5W (4)
Aug 60	Det	Agana	AD-5W (7), AD-5Q (7), WF-2 (1)
	Det C	CVA-19	WF-2 (3)
	Det F	CVA-34	AD-5Q (2)
	Det N	CVS-12	AD-5W (4)
Sep 60	Det	Agana	AD-5W (7), AD-5Q (7), WF-2 (1)
	Det C	CVA-19	WF-2 (3)
	Det F	CVA-34	AD-5Q (2)
	Det N	CVS-12	AD-5W (4)
Oct 60	Det	Agana	AD-5W (4), AD-5Q (7), WF-2 (4)
	Det C	CVA-19	WF-2 (3)
	Det D	CVA-43	WF-2 (3)
	Det F	CVA-34	AD-5Q (1)
	Det N	CVS-12	AD-5W (4)
Nov 60	Det	Agana	AD-5W (4), AD-5Q (8), WF-2 (3)
	Det C	CVA-19	WF-2 (3)
	Det D	CVA-43	WF-2 (3)
	Det N	CVS-12	AD-5W (4)
Dec 60	Det	Agana	AD-5W (4), AD-5Q (8), WF-2 (3)
	Det C	CVA-19	WF-2 (2)
	Det D	CVA-43	WF-2 (3)
Jan 61	Det	Agana	AD-5W (4), AD-5Q (9), WF-2 (3)
	Det C	CVA-19	WF-2 (3)
	Det D	CVA-43	WF-2 (3)
Feb 61	Det	Agana	AD-5Q (7), WF-2 (2)
	Det C	CVA-19	WF-2 (3)
	Det D	CVA-43	WF-2 (3)
Mar 61	Det	Agana	AD-5Q (6), WF-2 (5)
	Det D	CVA-43	AD-5Q (2), WF-2 (3)
Apr 61	Det	Agana	AD-5Q (8), WF-2 (5)
	Det D	CVA-43	WF-2 (3)
May 61	Det	Agana	AD-5Q (6), WF-2 (2)
	Det B	Atsugi	AD-5Q (2)
Jun 61	Det	Atsugi	AD-5Q (4)
	Det B	Atsugi	AD-5Q (2)
Jul 61	Det	Agana	AD-5Q (9), TF-1Q (1)
	Det A	Atsugi	AD-5Q (2)
	Det B	Cubi	AD-5Q (2)
	Det E	CVA-31	AD-5Q (2)
Aug 61		Alameda	AD-5Q (9), TF-1Q (2)
	Det A	Atsugi	AD-5Q (2)
	Det B	Cubi	AD-5Q (2)
	Det E	CVA-31	AD-5Q (2)
	Det M	CVA-61	AD-5Q (2)
Sep 61		Alameda	AD-5Q (9), TF-1Q (2)
	Det A	Alameda	AD-5Q (2)
	Det B	Cubi	AD-5Q (2)
	Det E	CVA-31	AD-5Q (2)
	Det M	Atsugi	AD-5Q (2)
Oct 61		Alameda	AD-5Q (11), TF-1Q (2)

Date	Det	Location	Aircraft
	Det B	Cubi	AD-5Q (2)
	Det E	CVA-31	AD-5Q (2)
	Det F	Alameda	AD-5Q (2)
	Det M	Atsugi	AD-5Q (2)
Nov 61		Alameda	AD-5Q (11), TF-1Q (2)
	Det B	CVA-14	AD-5Q (2)
	Det E	CVA-31	AD-5Q (2)
	Det F	CVA-16	AD-5Q (2)
	Det M	CVA-61	AD-5Q (1)
Dec 61		Alameda	AD-5Q (11), TF-1Q (2)
	Det B	CVA-14	AD-5Q (2)
	Det C	CVA-43	AD-5Q (2)
	Det F	CVA-16	AD-5Q (2)
	Det M	CVA-61	AD-5Q (1)
Jan 62		Alameda	AD-5Q (14), TF-1Q (2)
	Det F	CVA-16	AD-5Q (1)
	Det L	CVA-19	AD-5Q (2)
	Det M	CVA-61	AD-5Q (2)
Feb 62		Alameda	AD-5Q (16), TF-1Q (2)
	Det F	CVA-16	AD-5Q (2)
	Det L	Atsugi	AD-5Q (2)
	Det M	CVA-61	AD-5Q (1)
Mar 62		Alameda	AD-5Q (15), TF-1Q (2)
	Det F	Atsugi	AD-5Q (2)
	Det L	Cubi	AD-5Q (2)
Apr 62		Alameda	AD-5Q (13), TF-1Q (2)
	Det A	Atsugi	AD-5Q (2)
	Det F	CVA-16	AD-5Q (2)
	Det L	Cubi	AD-5Q (2)
May 62		Alameda	AD-5Q (13), TF-1Q (2)
	Det E	Cubi	AD-5Q (5)
Jun 62		Alameda	AD-5Q (17), TF-1Q (2)
	Det I	Cubi	AD-5Q (2)
Jul 62		Alameda	AD-5Q (16), TF-1Q (2)
	Det I	Cubi	AD-5Q (3)
Aug 62		Alameda	AD-5Q (15), TF-1Q (2)
	Det I	Cubi	AD-5Q (3)
Sep 62		Alameda	AD-5Q (16), TF-1Q (2)
	Det I	Cubi	AD-5Q (3)
Oct 62		Alameda	EA-1F (14), EC-1A (2)
	Det I	Cubi	EA-1F (4)
Nov 62		Alameda	EA-1F (14), EC-1A (1)
	Det F	Barb. Pt.	EC-1A (1)
	Det I	Cubi	EA-1F (4)
Dec 62		Alameda	EA-1F (12)
	Det F	Alameda	EA-1F (2)
	Det I	Cubi	EA-1F (3)
Jan 63		Alameda	EA-1F (11), EC-1A (1)
	Det F	Barb. Pt.	EA-1F (2)
	Det I	Cubi	EA-1F (4)
Feb 63		Alameda	EA-1F (12), EC-1A (2)
	Det F	Barb. Pt.	EA-1F (2)
	Det I	Cubi	EA-1F (4)
Mar 63		Alameda	EA-1F (11), EC-1A (2)
	Det F	Barb. Pt.	EA-1F (2)
	Det I	Cubi	EA-1F (3)
Apr 63		Alameda	EA-1F (12), EC-1A (2)
	Det I	Cubi	EA-1F (6)
May 63		Alameda	EA-1F (12), EC-1A (2)
	Det I	Cubi	EA-1F (6)
Jun 63		Alameda	EA-1F (12), EC-1A (2)
	Det I	Cubi	EA-1F (6)
Jul 63		Alameda	EA-1F (11), EC-1A (2)
	Det I	Cubi	EA-1F (6)
Aug 63		Alameda	EA-1F (9), EC-1A (2), A-1G (1)
	Det I	Cubi	EA-1F (6)
Sep 63		Alameda	EA-1F (10), EC-1A (2), A-1G (1)
	Det I	Cubi	EA-1F (7)
Oct 63		Alameda	EA-1F (11), EC-1A (2), A-1G (1)
	Det I	Cubi	EA-1F (7)
Nov 63		Alameda	EA-1F (11), EC-1A (2), A-1G (1)
	Det I	Cubi	EA-1F (7)
Dec 63		Alameda	EA-1F (11), EC-1A (2), A-1G (1)
	Det I	Cubi	EA-1F (7)
Jan 64		Alameda	EA-1F (11), EC-1A (2), A-1G (1)

Date	Base	Aircraft
Det I	Cubi	EA-1F (7)
Feb 64	Alameda	EA-1F (10), EC-1A (2), A-1G (1)
Det I	Cubi	EA-1F (8)
Mar 64	Alameda	EA-1F (10), EC-1A (2), A-1G (1)
Det I	Cubi	EA-1F (8)
Apr 64	Alameda	EA-1F (9), EC-1A (2), A-1G (1)
Det I	Cubi	EA-1F (6)
May 64	Alameda	EA-1F (12), EC-1A (1), A-1G (1)
Det I	Cubi	EA-1F (6)
Jun 64	Alameda	EA-1F (13), EC-1A (2), A-1G (1)
Det I	Cubi	EA-1F (6)
Jul 64	Alameda	EA-1F (11), EC-1A (1), A-1G (1)
Det I	Cubi	EA-1F (7)
Aug 64	Alameda	EA-1F (11), EC-1A (1), A-1G (1)
Det I	Cubi	EA-1F (7)
Sep 64	Alameda	EA-1F (11), EC-1A (2), A-1G (1)
Det I	Cubi	EA-1F (7)
Oct 64	Alameda	EA-1F (9), EC-1A (2), A-1G (1)
Det I	Cubi	EA-1F (7)
Nov 64	Alameda	EA-1F (10), EC-1A (2), A-1G (1)
Det I	Cubi	EA-1F (6)
Dec 64	Alameda	EA-1F (10), EC-1A (2), A-1G (1)
Det I	Cubi	EA-1F (6)
Jan 65	Alameda	EA-1F (11), EC-1A (2)
Det I	Cubi	EA-1F (6)
Feb 65	Alameda	EA-1F (12), EC-1A (2)
Det I	Cubi	EA-1F (7)
Mar 65	Alameda	EA-1F (10), EC-1A (2)
Det I	Cubi	EA-1F (7)
Apr 65	Alameda	EA-1F (10), EC-1A (2)
Det I	Cubi	EA-1F (7)
May 65	Alameda	EA-1F (7), EC-1A (2)
Det I	Cubi	EA-1F (6)
Jun 65	Alameda	EA-1F (6), A-1E (1), EC-1A (2)
Det I	Cubi	EA-1F (11)
Jul 65	Alameda	EA-1F (7), A-1E (3), EC-1A (2)
Det I	Cubi	EA-1F (11)
Aug 65	Alameda	EA-1F (7), A-1E (3), EC-1A (2)
Det I	Cubi	EA-1F (11)
Sep 65	Pt. Mugu	EA-1F (6), A-1E (1), EC-1A (2)
Det I	Cubi	EA-1F (10)
Oct 65	Alameda	EA-1F (5), A-1E (2), EC-1A (1)
Det I	Cubi	EA-1F (11)
Nov 65	Alameda	EA-1F (5), A-1E (2), EC-1A (2)
Det I	Cubi	EA-1F (10)
Dec 65	Alameda	EA-1F (7), A-1E (2), EC-1A (2)
Det I	Cubi	EA-1F (8)
Jan 66	Alameda	EA-1F (7), A-1E (3), EC-1A (1)
Det I	Cubi	EA-1F (9)
Feb 66	Alameda	EA-1F (7), A-1E (2), EC-1A (2)
Mar 66	Alameda	EA-1F (7), A-1E (2), EC-1A (2)
Apr 66	Alameda	EA-1F (19), A-1E (2), EC-1A (2)
May 66	Alameda	EA-1F (7), A-1E (2), EC-1A (2)
Jun 66	Alameda	EA-1F (7), A-1E (2), EC-1A (2)
Jul 66	Alameda	EA-1F (10), A-1E (2), EC-1A (2)
Det I	Cubi	EA-1F (1)
Aug 66	Alameda	EA-1F (11), A-1E (1), EC-1A (2)
Det I	Cubi	EA-1F (6)
Sep 66	Alameda	EA-1F (11), A-1E (1), EC-1A (2)
Det I	Cubi	EA-1F (10)
Oct 66	Alameda	EA-1F (11), A-1E (1), EC-1A (2)
Det I	Cubi	EA-1F (10)
Nov 66	Alameda	EA-1F (9), EC-1A (2)
Det I	Cubi	EA-1F (13)
Dec 66	Alameda	EA-1F (9), EC-1A (2), C-1A (1)
Det I	Cubi	EA-1F (12)
Jan 67	Alameda	EA-1F (9), EC-1A (2), C-1A (1)
Det I	Cubi	EA-1F (12)
Feb 67	Alameda	EA-1F (10), EC-1A (2)
Det I	Cubi	EA-1F (12)
Mar 67	Alameda	EA-1F (10), EC-1A (2)
Det I	Cubi	EA-1F (12)
Apr 67	Alameda	EA-1F (10), A-1E (1), EC-1A (1)
Det I	Cubi	EA-1F (11)
May 67	Alameda	EA-1F (10), A-1E (1), EC-1A (2)
Det I	Cubi	EA-1F (12)
Jun 67	Alameda	EA-1F (9), A-1E (1), EC-1A (2), EKA-3B (1)
Det I	Cubi	EA-1F (10)
Jul 67	Alameda	EA-1F (11), A-1E (2), EC-1A (2), EKA-3B (1)
Det I	Cubi	EA-1F (8)
Aug 67	Alameda	EA-1F (11), A-1E (2), EC-1A (2), EKA-3B (1)
Det I	Cubi	EA-1F (8)
Sep 67	Alameda	A-1E (1), EC-1A (2), EKA-3B (2), KA-3B (2)
Det I	Cubi	EA-1F (7)
Oct 67	Alameda	A-1E (1), EC-1A (2), EKA-3B (2), KA-3B (2)
Det I	Cubi	EA-1F (7)
Nov 67	Alameda	EA-1F (4), EKA-3B (4), KA-3B (2)
		+ 1 EKA-3B det
Det 63	Alameda	EA-1F (3)
Det 101	Cubi	EA-1F (8)
Dec 67	CVA-63	A-1E (1), EKA-3B (5), KA-3B (1)
		+ 2 EKA-3B dets
Det 63	Alameda	EA-1F (3)
Det 101	Cubi	EA-1F (7)
Jan 68	CVA-63	A-1E (1), EKA-3B (5), KA-3B (1)
		+ 3 EKA-3B dets
Det 63	Alameda	EA-1F (3)
Det 101	Cubi	EA-1F (6)
Feb 68	CVA-63	A-1E (1), EA-1F (2), EKA-3B (5), KA-3B (1)
		+ 3 EKA-3B dets
Det 63	Alameda	EA-1F (3)
Det 101	Cubi	EA-1F (6)
Mar 68	CVA-63	A-1E (1), EA-1F (2), EKA-3B (6)
		+ 5 EKA-3B dets
Det 63	Alameda	EA-1F (3)
Apr 68	CVA-63	A-1E (1), EA-1F (6), EKA-3B (6)
		+ 5 EKA-3B dets
Det 63	Alameda	EA-1F (3)
May 68	CVA-63	A-1E (1), EA-1F (3), EKA-3B (8), KA-3B (1)
		+ 5 EKA-3B dets
Jun 68	CVA-63	A-1E (1), EA-1F (3), EKA-3B (9)
		+ 5 EKA-3B dets
Jul 68	CVA-63	A-1E (1), EKA-3B (9)
		+ 5 EKA-3B dets
Aug 68	CVA-63	A-1E (1), EKA-3B (6)
		+ 5 EKA-3B dets
Sep 68	Alameda	A-1E (1)
		+ 7 EKA-3B dets
VAQ-130:		
Oct 68	CVA-63	A-1E (1), EKA-3B (4)
		+ 4 EKA-3B dets
Nov 68	CVA-63	A-1E (1), EA-1F (6), EKA-3B (6)
		+ 6 EKA-3B dets
Det 63	Alameda	EA-1F (3)
Dec 68	Alameda	A-1E (1), EKA-3B (4), KA-3B (3)
		+ 6 EKA-3B dets

ations at Cubi Pt. from which to prosecute this mission. It became Detachment 1 on 14 May and drew its assets from Det F (CVA-16), Det D (CVA-43), Det A (CVA-41), and Det L (CVA-19). Over the next two years VAW-13 alternated operations from Tan Son Nhut Air Base near Saigon with the Air Force F-102s. After the cessation of the night fighter mission in Vietnam, Det 1 Cubi continued to provide jamming services for ships and fighters while operating from land bases in Vietnam, Korea, Okinawa, and Japan.

In March 1965, Det 1 Cubi Pt. began supplying EA-1F TAD Dets (usually two EA-1Fs) to carriers on Yankee Station. While ships rotated on-and-off Yankee Station, the squadron's Skyraiders would deck hop from the retiring carrier to the reporting one. Assignments were on

an as-needed basis and could last anywhere from a few days to up to a month.

For operations off Vietnam, the EA-1F carried a 300 gal centerline fuel tank, an APS-31 search radar on the right inner wing pylon and two ALT-2 jammers on the outermost outer wing pylons, and an ALE-2 chaff dispenser on the left inner wing pylon. The mission was conducted by flying orbits between 8,000 to 10,000 ft and the primary targets were AAA Whiff and Fire Can systems. On rare occasions, if the stars were aligned, the SA-2 Fan Song fire control system could also be jammed. The EA-1Fs were fitted with two 20mm cannon and were occasionally called upon to supply SAR support. On one such occasion, the squadron lost its first and last EA-1F to enemy action. EA-1F, BuNo 132540, was crewed by LTJG MD McMican, LTJG Gerald Romano, PO2 Tom Plants, and PO3

Bill Amspacher, when on 2 June 1966 they were diverted to the shoot-down site of a VA-23 A-4E. All four were killed in action after being hit by ground fire. Three operational aircraft losses followed, the first on 20 June 1966. LT John McDonough was killed in BuNo 135010 due to a cold cat shot off Hancock while his crew of two escaped. On 10 September, BuNo 132543 from CVA-42 was downed at night by an electrical fire. A successful ditching was made and all four crewmembers, LT Lanny Cox, LTJG Robert Carlton, AD1 Andy Anderson, and ATN2 Gordon Johnson, were rescued by an Air Force HU-16. The third EA-1F, BuNo 133770, went down due to an engine failure while on a ferry flight from Cubi Pt. on 25 September 1967. All crewmen were rescued. Then on 14 February 1968, a VAW-13 EA-1F on a ferry flight from Cubi to the Tonkin Gulf in company with a VA-25 A-1H was jumped by Chinese F.9s (MiG-19s) off Hainan

Island. The EA-1F escaped, but the A-1H did not.

Meanwhile, on the home front, the Navy was tooling up the EA-1F's replacement, the EKA-3B. The squadron's first EKA-3B was acquired in June 1967 and VAW-13 was redesignated VAQ-130 on 1 October 1968. Det 63 made VAW-13's last EA-1F combat cruise from

---

Below, VAW-13 AD-5Q BuNo 135010 was assigned to CVG-2 aboard the USS Midway (CVA-41) and is seen here landing at NAF Atsugi, Japan, on 1 August 1961. The outer wing pylons have two ALT-2 jammer pods, the right wing inner pylon has an APS-31 radar pod mounted, the centerline carries a 300 gal fuel tank, and the left inner wing pylon has an ALE-2 chaff dispenser mounted on it. (Toyokazu Matsuzaki

November 1967 to June 1968 from the USS Kitty Hawk. The squadron's last Skyraider, an A-1E, was retired in January 1969.

Above, VAW-13 EA-1F landing aboard the USS Constellation CVA-64 in the Gulf of Tonkln on 14 September 1967. Fin stripe was blue. (USN) Below, the last Skyraider used by the squadron was A-1E BuNo 132446 and the only aircraft to carry the VAQ-130 designation and the only one with double nuts cowl numbers. (William Swisher collection)

ATTACK SQUADRON FIFTEEN, VA-15 "VALIONS"

VA-15 was first established as Torpedo Squadron Four (VT-4) on 10 January 1942 aboard the USS Ranger (CV-4) at Grassy Bay, Bermuda, with TBD-1 Devastators. In August 1942, the TBDs were replaced with TBF-1 Avengers and by the end of 1946 the squadron had operated six different types of TBF/TBMs. On 15 November 1946, VT-4 was redesignated Attack Squadron Two A (VA-2A) while flying 15 TBM-3Es and 5 TBM-3Qs. VA-2A became Attack Squadron Fifteen (VA-15) on 2 August 1948.

In August 1949, the squadron received eleven AD-4s while operating twelve TBM-3Es at NAS Cecil Field, FL. By October 1949, VA-15 had a full complement of sixteen AD-4s which then fluctuated between seventeen and nineteen aircraft through 22 May 1950 when all squadrons in CVG-1 were designated training squadrons. VA-15 was responsible for training attack pilots in glide bombing, dive-bombing, rocket firing, day-and-night tactics and carrier qualifications in the AD Skyraider. This assignment lasted less than a year.

The first Skyraider deployment was the 12 September 1949 shakedown cruise to the Caribbean aboard the USS Philippine Sea (CV-47). The second was again aboard CV-47 to the North Atlantic from 14 February 1950 for exercise Portex I in concert with CVB-43 and CV-32. This was followed by Operation CARIBEX in the Caribbean on 16 March involving the Vieques Island gunnery and bombing range.

In May 1950, during VF-172 F2H-

Above, VA-15 AD-4s aboard the Wasp for Operation Mainbrace in 1952. (USN) Below, VA-15 AD-6 BuNo 135236 on 9 June 1954 at NAS Cecil Field. Fin tip was green. (Ginter collection)

2 CarQuals an incoming Banshee lost its hook point and hit the pack of aircraft damaging two F8Fs and a VA-15 AD as well as knocking another overboard.

CVG-1 deploying aboard the USS Coral Sea (CVB-43) from 20 March through 6 October 1951 on a cruise to the Mediterranean with AD-4/4Ls.

CVG-1 and VA-15 took part in the first NATO naval operation, Operation Mainbrace, conducted in the North Atlantic from 13 to 23 September 1952 while deployed aboard the USS Wasp (CV-18) from 24 May through 11 October 1952.

From 11 June through 3 December 1953, VA-15 deployed aboard the USS Franklin D. Roosevelt (CVA-42) with AD-4/4B/4Ls. During this period the squadron received the coveted Battle "E" award. While operating in the Med, LTJG Charles MacDowell, lost his engine enroute to a target range in Afyon, Turkey. He made a dead-stick, gear-up landing on a 3,000ft plateau and was welcomed by a camel caravan. He was then transported to Afyon where the CAG was notified and a plan was devised to rescue the aircraft. With the aid of a Turkish Air Force captain, a caravan of rescue and tow equipment was sent to retrieve the AD and transport

Above left, VA-15 AD-6 in flight from the Midway in 1954. (USN) At left, VA-15 AD-6 Skyraider T/513 traps aboard the USS Midway (CVA-41) in 1954. Fin tip was green and the trailing edge of the rudder was white. (USN) Below, VA-15 AD-6 BuNo 139782 on 5 March 1957 with 400 series modex numbers as used while aboard Forrestal. Fin tip was orange. (Doug Olson)

it the 60-miles to Afyon. It took five days to get to Afyon where a repair team from FASRON-77 replaced the engine and repaired any landing damage. Once repaired, MacDowell flew to Athens, Greece; Naples; Crete and finally the FDR.

A World Cruise from 27 December 1954 through 14 July 1955 was conducted aboard the USS Midway (CVA-41) with AD-6s. In February 1955, the squadron supported the evacuation of Chinese National civilians and military personnel from the Tachen Islands during their bombardment by the People's Republic of China.

In September 1956, VA-15 conducted carrier qualifications aboard CVA-59. This was followed by three deployments aboard the USS Forrestal (CVA-59). The first was a shakedown cruise to the Azores from 7 November through 12 December 1956 during the Suez Crisis. The second Med cruise was from 15 January through 22 July 1957 with AD-6s. Then it was off to the North Atlantic from 16 August through 21 October 1957 where VA-15 took part in Operation Seaspray and NATO Operation Strike Back.

In June 1958, the squadron received an AD-5 and in August, VA-

---

Below, VA-15 (T/4XX), VAW-12 (NE/7XX), and VA(AW)-33 (SS/8XX) Skyraiders aboard the USS Forrestal (CVA-59) in 1957. (USN)

Above, VA-15 Skyraiders run up prior to a practice mission in 1960 aboard CVA-42. (USN) Below, seven VA-15 AD-6s aboard the USS Forrestal (CVA-59) with five VAW-33 AD-5s in September 1957. (National Archives)

---

15 was assigned the additional mission of inflight refueling. In September through November, the squadron participated in a series of operations off the East Coast while assigned to the FDR.

VA-15 returned to the USS Franklin D. Roosevelt in 1959 and 1960 for two deployments to the Med. The first was from 13 February through 1 September 1959. During this cruise, VA-15 took part in Operation Big Deal (25-28 February),

Operation Tuner-Up (23-24 March), Operation Top Weight (13-17 April), and Operation Green Swing (9-23 April). During Operation Top Weight, VA-15 conducted simulated nuclear strikes against Spain, Italy, Greece, and Turkey. During Operation Green Swing they provided close air support for a simulated amphibious landing on the coast of Italy. In operations off the United States, the squadron participated in LANTFLEX 3-59 From 19-29 October and in WEXUAL 8 & 9. In January 1960, VA-15 took part in LANTFLEX 1-60 with simulated nuclear strikes on America.

The second FDR cruise was from 28 January to 24 August 1960 during which a Battle "E" award was bestowed on the squadron on 1 July 1960. During the cruise, the squadron participated in Operations Big Deal, Quicktrain, Haystrike, Rejex, Royal

Flush, and Purple Sage.

The squadron was next transferred to CVG-10 and the USS Shangri-La (CVA-38) for a short two week deployment from 14 to 28 November 1960 to the Caribbean to guard against possible infiltration of insurgents into Guatemala and Nicaragua by Cuban backed guerrillas.

In 1961, VA-15 returned to CVA-42 and CVG-1 for two more deployments with the Skyraider. Carrier qualifications and refresher training was conducted off Guantanamo from 7 to 22 January 1961 prior to deploying to the Med. During the first cruise to the Med from 15 February to 28 August 1961, the squadron received its third Battle "E" award on 1 July 1961.

The second deployment was in the vicinity of the Dominican Republic from 19 through 30 November 1961 to support the newly established government.

The squadron also participated in the Caribbean shakedown cruise of the nuclear powered USS Enterprise (CVAN-65) from 5 February through 8 April 1962.

The squadron returned to FDR for a Joint Chief of Staff Orientation Cruise from 20 April to 18 May 1962. Further carrier qualifications were made aboard CVA-42 in May, June, July and August

The next Med cruise aboard CVA-42 was from 14 September 1962 to 22 April 1963, once again to the Med with the AD-6/A-1H. In May, an emergency deployment was made to Haiti to discourage a rebel coup.

On 20 December 1963, CVG-1 was redesignated CVW-1 prior to VA-15's final Skyraider cruise from 28 April to 22 December 1964. Workups for the deployment were conducted out of Guantanamo from 21 December 1963 to 10 January 1964. During the Med cruise, VA-15 took part in Operations FALLEX 64, and Poopdeck IV.

Transition to the Skyhawk began in late April 1965 and by December the squadron was completely transitioned to the A-4B. VA-15 made three deployments with A-4B/Cs before it was disestablished on 1 June 1969.

Above, VA-15 A-1H BuNo 139776 landing at sea aboard CVA-59 during Operation Strikeback in October 1957. (USN) Below, VA-15 AD-6 BuNo 139675 on 19 July 1962. (Ginter collection) Bottom, VA-15 AD-6 landing aboard the USS Enterprise (CVAN-65) in 1962. (USN)

ATTACK SQUADRON SIXTEEN, VA-16

Attack Squadron Sixteen (VA-16) was established at NAS Oceana, VA, on 1 June 1955 along with Air Task Group One Eighty Two (ATG-182). The squadron was under the command of CDR Bartholomew J. Connolly III and equipped with AD-6s while assigned to the USS Lake Champlain (CVA-39). VA-16 deployed to the Med from 21 January to 27 July 1957 where in April they operated off the coast of Lebanon during the Jordanian crisis. On 18 December 1957, VA-16 became the first fleet squadron to conduct air-to-air tanking using the Douglas "buddy store". The first transfer was over Oceana to an F9F-8. This was followed on 9 January 1958 by VA-16 becoming the first carrier based AD squadron to conduct in-flight refueling. The squadron was temporarily deployed aboard the USS Ranger (CVA-61) at the time. Two months later, on 1 March 1958, VA-16 was disestab-

Above, VA-16 AD-6 BuNo 139724 aboard CVA-39 in 1956. (USN via Angelo Romano) Below, CVA-39 in 1957 with VA-16 AD-6s on the after deck. (USN via Angelo Romano)

lished while under the command of CDR Richard W. Willis.

At left, VA-16 AD-6 ran off the runway at NAS Oceana, VA, and came to rest in shallow water in November 1957. VA-16's tail code was changed from "O" to "AN" on 1 July 1957. (USN) Below, VA-16 AD-6s BuNos 139763 (O/408), 139755 (O/413), 139765 (O/402), 139726 (O/409), 139758 (O/405), and 139754 (O/754) in formation in early 1957. Trim was in orange. (USAF)

ATTACK SQUADRON NINETEEN A, VA-19A
ATTACK SQUADRON ONE - NINETY - FOUR, VA-194

VA-194 was originally established as Bombing Squadron Nineteen (VB-19) on 15 August 1943 at NAAS Los Alamitos, CA. The squadron's first aircraft was an SBD-5 which was replaced by the SB2C-1 in April 1944. VB-19 was redesignated Attack Squadron Nineteen A (VA-19A) at NAS Alameda, CA, on 15 November 1946. VB-19/VA-19A flew six versions of the Helldiver before acquiring the AD-1 in February 1947. VA-19A was the first fleet squadron to equip with the Skyraider. By 21 February there were thirteen AD-1s and seven SB2C-5s on-hand. The squadron was assigned to CVG-19 and initially the USS Antietam (CV-36). Carrier qualifications and short port-to-port calls were made up-and-down the West Coast until the squadron was re-assigned to the

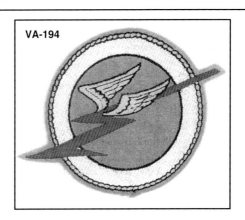

VA-194

USS Boxer (CV-21) in April and by June had a full complement of twenty AD-1s. West Coast operations continued including operations aboard the USS Princeton (CV-37).

In March 1948, an AD-1 returning from a dive-bombing qualification flight couldn't lower its tailwheel. After

numerous attempts to force it down and with fuel remaining at under 50 gal, the pilot elected to lower the belly speed/dive brake and use it as a tail wheel. He came in at 85 knots and made an excellent landing. Damage was limited to the aft trailing edge of

the dive brake and after its replace-ment the AD-1 was back in the air the next day.

VA-19A was redesignated VA-194 on 24 August 1948 and was dis-established on 1 December 1949.

Above, VA-194 AD-1 deck launches from the USS Princeton (CV-37) on 12 August 1948. (USN) At right, AD-1 (B/320) at Alameda with an APS-4 radar pod under the left wing. (NMNA) Below, VA-19A and VA-20A AD-1s on the after deck of CV-21. (NMNA) At left, VA-19A AD-1s prepare to launch from the USS Boxer (CV-21) in 1947-1948. (USN via Tailhook)

ATTACK SQUADRON TWENTY A, VA-20A "TIGERS"
ATTACK SQUADRON ONE - NINE - FIVE, VA-195 "DAMBUSTERS"

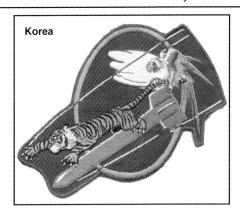

Korea

1955

VA-195 was originally established as Torpedo Squadron Nineteen (VT-19) on 15 August 1943 at NAS Los Alamitos, CA. They were equipped with TBM-1s and operated four other versions of the Avenger before receiving AD-1s in June 1947. The unit was redesignated Attack Squadron Twenty A (VA-20A) on 15 November 1946 at NAS Moffett Field, CA, and became VA-195 on 24 August 1948.

Like VA-19A, VA-20A Skyraiders were assigned to CVG-19 and to the USS Boxer (CV-21). VA-20A initially operated twenty AD-1s, too, just like its sister squadron. A AD-1Q was also operated in July and August 1948 and sixteen AD-2s replaced an equal number of AD-1s in August. In January 1949, the AD-2s were replaced with sixteen AD-3s. In January 1950, four AD-2Qs were acquired to operate alongside the unit's sixteen AD-3s for the squadron's first deployment. The Boxer departed San Diego headed

for the Western Pacific on 11 January 1950 and returned to Alameda on 13 June 1950. Ports-of-call were: Pearl Harbor, Yokosuka, Hong Kong, Subic Bay, Manila, Sasebo, Inchon, Seoul, and Singapore. Joint operations were also conducted with British carriers, which helped prepare the two Navies for their combined operations during the imminent Korean War.

Due to the onset of the Korean War, the air group made a quick turnaround and was reassigned to the USS Princeton (CV-37). After upgrad-

ing to the AD-4, the squadron sailed for Korea on 9 November 1950 with twelve AD-4s and one AD-4Q. For the deployment, which ended on 29 May 1951, LCDR H.D. Carlson was in

Below, VA-20A AD-1 BuNo 09204 and VF-20A F8F-1 BuNo 95318 in flight over San Francisco on 2 June 1947. Rudder trim was white. (William T. Larkins)

Above, VA-20A AD-1 BuNo 09199 unfolds wings at NAS Alameda, CA, on 2 June 1947. (William T. Larkins) Below, VA-20A AD-1's gear collapsed and B/405 slid into the 5" gun mount and over the side on 17 September 1947 aboard CV-21. (USN via Tailhook)

After VA-20A was redesignated VA-195 on 24 August 1948, their aircraft numbers were changed from the 400 series to the 500 series. Above, VA-195 AD-3 BuNo 122750 taxis forward past manned F8Fs aboard Boxer in early 1950. (USN) Below, 122750 piloted by ENS Robert Bennett takes the barrier on CV-21 in early 1950. Fin tip was green. (USN)

Above, VA-195 AD-2Q BuNo 122377 from CV-21 in early 1950. VA-195 Skyraiders had a green fin tip and prop hub. (NMNA) At right, squadron Skyraider in slow flight with gear and fuselage dive brakes open. (USN) Bottom, VA-195 AD-3 BuNo 122724 at NAS Alameda, CA, in 1949. Fin tip was green. (William T. Larkins)

command. The ship stayed on station and CVG-19 was relieved by CVG-19X) and VA-195's AD-s were transferred to VA-55 while the officers were flown back to CONUS and the enlisted men were transported by ship.

VA-195 flew its first combat mission on 5 December 1950 in support of Marines near the Chosen Reservoir. The remainder of December had the Princeton covering the withdrawal of Hungnam with the other three fast carriers: CV-32, CV-45 and CV-47. During daylight hours, VA-195 was tasked with providing close support inside the embarkation area and of interdicting enemy supply lines outside the embarkation area.

CARLSON'S CANYON & THE BRIDGES OF TOKO-RI

On 2 March 1951, VF-193's skipper, LCDR C.M. Craig, was returning from a strike on the Kilchu bridges in his Corsair when he discovered two rail bridges under construction south of Kilchu. One bridge appeared complete with five concrete support abutments and steel spans; the other also had five concrete abutments but lacked the steel spans. A hasty attack was conducted that afternoon with ineffective results to the bridges' approaches. The next day, VA-195 returned with a well-planned attack prosecuted by eight AD-4s equipped

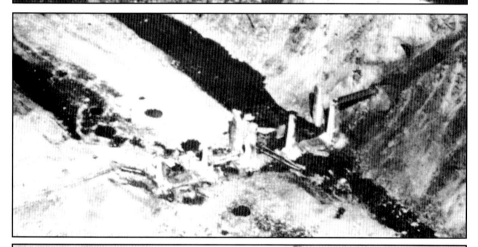

with three 2,000 lb bombs. The squadron CO, "Swede" Carlson, led the attack which dropped one span and damaged three others. Upon return to the ship, the Task Force commander, RADM Ralph Ofstie, dubbed the bridgehead "Carlson's Canyon". The battle for Carlson's Canyon was not over! The North Koreans would make repairs and we would attack again over-and-over. The effort became immortalized by James Michener's fact-based fictional book, the Bridges of Toko-Ri, and the blockbuster movie that followed staring Richard Holden.

On 15 March 1951, the bridges were struck again, this time with napalm added to burn the supporting wood cribbing. The results left three spans standing. Throughout March the bridge-busting effort expanded north and south of Carlson's Canyon and included tunnel-busting. Carlson's Canyon was struck again on 27 March by an Air Force B-29 with long-delay action bombs. However, communist repairs prompted another TF-77 attack on 2 April. Two raids were conducted and all spans were dropped after which the North Koreans gave up trying to repair the bridges.

HWACHON DAM BUSTERS

On 9 April 1951, as the 8th Army moved northward, the North Koreans opened the floodgates on the Hwachon Dam, overflowing the Pukhan River, knocking out bridges and temporarily halting the Army's

Top-to-bottom, bombed-up VA-195 AD-4 BuNo 123937 inbound to a Korean target. Note Dambuster insignia below canopy. (Paul Minert collection) The rail bridges in Carlson's Canyon after the 15 March 1951 strike with three out of six spans still standing. (USN) 2 April 1951 view of the bridges after the day's two strikes with all spans down. (USN) VA-195 AD-4 loaded with three 2,000 lb bombs warms up prior to the 3 March 1951 strike on the bridges. (USN)

Above, LCDR Carlson, CO VA-195. Above right, VA-195 AD-4 from CV-37 enroute to torpedo the Hwachon Dam on 1 May 1951. (NMNA) Below right, VA-195 AD-4 lowering wings and taxiing forward prior to launch on 1 May 1951. (NMNA) Bottom, VA-195 Skyraider deck launches with a torpedo and two napalm tanks for an attack on the Hwachon Dam on 1 May 1951. (via Tailhook)

advance. In response, the Army attacked the dam on 11 April with a small group of Rangers and mechanized cavalry, but failed to capture or incapacitate it. During a second attempt on 21 April, they tried again and secured the dam briefly before an enemy offensive re-secured it. Again the Army began exploring avenues to cripple the dam. Guided bombs launched from B-29s on 30 April were the first attempt. After the attempt failed, the Army requested a Naval strike from RADM Ofstie. VA-195 was then tasked with the mission and attacked with 2,000 lb bombs and 11.75" Tiny Tim rockets that afternoon. The Tiny Tims bounced off and the 2,000 lb bombs just chipped at

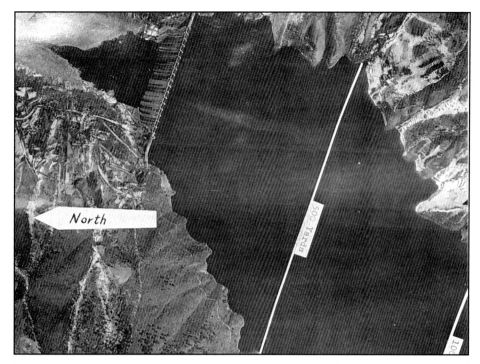

North

the concrete edges of the 240 ft thick dam that were fortified with rocks. So, like the Carlson Canyon raid, a well planed attack was scheduled for the next day, but not with bombs, with torpedoes. The strike force was made up of the CAG, CDR Richard Merrick in his AD-4Q and three VC-35 pilots flying AD-4Ns (LT Atlee Clapp, LT Frank Metzner, and LT Addison English) in the first division. VA-195's LCDR Carlson, ENS Robert Bennett, LTJG E.V. Phillips, and LTJG J.R. Sanderson made up the second division. The strike included four pilots who had dropped torpedoes (LCDR Carlson and the three VC-35 pilots), at least in practice, while the other five had not. Adding to the uncertainty of success was the fact that the 12 Mk-13 torpedoes aboard Princeton dated back to 1942-43 and that few if any ordnancemen aboard ship had ever touched a torpedo. In addition to the centerline mounted torpedo, most Skyraiders and the covering fighters carried two napalm tanks for post strike targets of opportunity. Luckily, almost no flak or AAA fire was encountered and the pilots were able to maintain the precise speeds and altitude needed for a successful drop. The attacks were made in flights of two line abreast and all eight torpedoes were launched. Six ran true and hit the dam, one ran erratically and one failed to explode. The strike totally destroyed one floodgate and blew away the bottom portion of another expelling water into the river below and thereby denying the enemy control of the dam's water system. Additionally, the dam's western abutment was holed. This would be the US Navy's last combat usage of an aerial torpedo and as a result of the action, VA-195 became known as the "Dambusters".

At top left, pre-strike dam intelligence photo with 500 and 1,000 yard marker lines added for pilot briefing. (USN) Middle left, Mk-13 torpedo impacts a floodgate near the center of the dam. (USN) At left, view of the river side of the dam when one torpedo hits near the center and another explodes on the western abutment. (USN)

Upon return to CONUS, the squadron reformed in July 1951, initially flying AD-1/-1Qs as interim aircraft. The Princeton didn't return to the states until 27 August and entered overhaul at the Puget Sound Naval Shipyard. She re-entered service and sailed to Korea arriving on 30 April 1952 for her second war deployment with CVG-19 and VA-195. For this deployment, CDR N.A. MacKinnon was in command and the squadron was equipped with seventeen AD-4s. On line periods were: 30 April, 1-13 May, 4-6 June, 6 July to 3 August, 18 August to 20 September, and from 4-16 October. Interdiction strikes were conducted against ground installations, bridges, storage depots, and the Suiho Dam. During June, two AD-4Ls were received as replacement aircraft. CVA-37 and VA-195 returned to Alameda on 3 November 1952 and the squadron was transferred to NAS Moffett Field, CA.

On the morning of 30 April 1952,

Above, LTJG Austin inspects the kitchen sink strapped to a 1,000 lb bomb he dropped on the communist capitol. (NavAirHist) Below, VA-195 AD-4 123820 readies for a mission over Korea in 1952. The pilot was LCDR E.V. Davidson, the squadron XO. Aircraft had 65 mission marks on the fuselage side. (USN)

just south of Wonsan, LTJG Ed Phillips lost the forward portion of his fin tip and rudder to a 40mm AA round. He made it back to the boat safely and the aircraft was repaired.

In June 1951, VA-195 took part in a thirty-five plane Skyraider raid with ADs from VA-65 and VA-115 on North Korean hydroelectric facilities. Upon return, LCDR M.K. Dennis commented, "We dropped everything on them but the kitchen sink." ADC R.B. Deland heard the remark and got with ADC H.J. Burdett and the maintenance crew figured out how to do it! The sink was bolted to a bomb and loaded onto LTJG Carl Austin's AD-4 along with an otherwise complete bomb load. The kitchen sink was then dropped on the North Korean capitol without incident.

On 28 July 1952, Princeton's Corsairs and Skyraiders attacked and destroyed a magnesite plant at Kilehu. Then, on 20 August, the North Korean supply center at Chagp'yong-

Above left, LTJG Edward Phillips's AD-4 BuNo 123933 was hit by a 40mm AA round south of Wonsan on 30 April 1951 but returned to the boat safely. (USN) At left, Phillips stands on the horizontal stabilizer to pose with his damaged vertical stabilizer. (Naval Aviation History) Below, VA-195 AD-6 BuNo 135252 at Oakland on 26 September 1954. (Swisher collection)

90

ni was struck by aircraft from Princeton and Essex teamed with Air Force F-84s. But the biggest Navy strike of the war occurred on 1 September 1952 with 142 aircraft from the Princeton, Boxer and Essex. They attacked and destroyed the Aoji oil refinery eight miles from the Soviet border out of range of Air Force strike aircraft.

The squadron's next two deployments were to the Western Pacific aboard the USS Oriskany (CVA-34). The first deployment with six AD-4Bs and ten AD-4NAs was from 14 September 1953 to 22 April 1954. The second deployment with sixteen AD-6s was from 2 March to 21 September 1955.

Two more AD-6 Skyraider deployments were made by VA-195. The first was aboard the USS Yorktown (CVA-10) from 9 March to 25 August 1957 with fourteen AD-6s and an AD-4Q. The last squadron deployment of the Skyraider was

aboard the USS Bon Homme Richard (CVA-31) from 1 November 1958 to 18 June 1959 with fourteen AD-6s and one AD-5. By 30 June, all but one AD-6 had been transferred out as the squadron began its transition to the A4D-2 Skyhawk.

Above, VA-195 AD-6 BuNo 135330 from the USS Oriskany (CVA-34) in 1955. Fin and wing tips were green and white candy stripes. (USN) Bottom, VA-195 AD-6 at right survives a Cougar barrier landing on the deck of CVA-34 in 1955. Note Cougar pilot still in the cockpit and blow-in doors open indicating a still running engine on the Cougar. Did he taxi forward with the barrier or did the barrier stop him between the Skyraider and Big Banjo? Fin and wing tips were green and white stripes. (USN)

Above, VA-195 AD-6 (B/514) taxis forward on CVA-10 in 1957 with a green diamond on the tail and green and white wing tips. (USN) At top right, VA-195 AD-6 traps aboard the USS Bon Homme Richard (CVA-31) in 1958. Note stylized "NM" tail code inside the green diamond on the tail. (USN) At right, VA-195 AD-6 BuNo 139675 low and slow with gear and hook down on 20 January 1959. (Jim Sullivan collection) Bottom right, VA-195 AD-6 BuNo 137629 at NAF Litchfield Park, AZ, on 17 April 1961 (Doug Olson) Below, loading rockets on B/504 and B-506 for weapons training off Hawaii. The one-day operation from CVA-10 saw VA-195 expend 120 rockets, 28,000 lbs of bombs, 16 napalm bombs, and 6,400 rounds of 20mm. (USN) Bottom, three VA-195 AD-6s, BuNos 139669 (B/406), 139670 (B/505) and 139661 (B/501) wait for their turn at the cat in 1957 on CVA-10. (USN)

COMPOSITE SQUADRON THIRTY-THREE, VC-33 "NIGHT HAWKS"
ALL WEATHER ATTACK SQUADRON THIRTY-THREE, VA(AW)-33
CARRIER AIRBORNE EARLY WARNING SQUADRON THIRTY-THREE, VAW-33
ELECTRONIC COUNTERMEASURES SQUADRON THIRTY-THREE, VAQ-33

Composite Squadron Thirty-Three (VC-33) was established on 13 May 1949 at NAS Norfolk, VA. As an anti-submarine unit, it was equipped with TBM-3E/-3Q/-3Ns. In May 1950, in preparation for a mission change, VC-33 received its first Skyraiders. By the 31st the squadron had eight AD-4Ns, ten TBM-3Es, and two TBM-3Ns on strength. On 5 June, the unit transferred to NAS Atlantic City, NJ, the home of VC-4, the then Atlantic Fleet night fighter/night attack squadron. With the move came the change in mission. VC-4 became AirLant's night fighter unit and VC-33 became AirLant's night attack unit. The change in mission was prophetic, as on 25 June the North Koreans invaded South Korea. The squadron would eventually send three detachments to the war zone. By the end of June, squadron strength had grown to eight AD-4Ns, three AD-4Qs, one AD-3N, three AD-2Qs, two AD-1Qs and two TBM-3Ns.

The first VC-33 Det to see combat was Det 3 aboard the USS Leyte (CV-32). The Officer-in-Charge (OIC) was LCDR Fred Silverthorn who had six pilots, two AD-3Ns and two AD-3Qs. The Leyte and CVG-3 originally deployed for a cruise to the Med beginning on 2 May 1950. Due to the escalating war in Korea, the ship was ordered on 20 August, while the crew

VC-33   1949

VC-33   VAW-33

was on liberty in Beirut, to report to the Pacific for duty off Korea. Sailing on 6 September from Norfolk, it transited the Panama Canal and arrived at San Diego on 18 September. They sailed again on 19 September and arrived off Korea on 3 October.

VC-33's first combat mission occurred on 9 October when the unit took part in an anti-mine exercise in Wonson Harbor. Two lines of CVG-3 F4Us and ADs flew across the harbor at a pre-described height and dropped hydrostatically fused bombs in an attempt to detonate any mines, but none exploded. After this useless mission, VC-33 settled into single aircraft night heckling (search-and-destroy) missions before covering the Chosin and Hungman withdrawals in

November and December. After Christmas 1950, the AD-3N/4Ns and AD-3Q/4Qs equipment and ability to carry two passengers/observers was utilized during daytime strikes. The AD-3Q was used as the Joint Operational Centers (JOC) radio relay aircraft. The pilot would orbit out of antiaircraft range in a "Dog" pattern and direct the strike. The AD-3Ns were utilized as pathfinders and would join in on the attack with rockets, bombs, guns, and napalm. VC-33

Below, VC-33 flightline at NAS Norfolk, VA, in 1950 with AD-2Q BuNo 122383 in the foreground. (Jim Sullivan collection)

aircraft were also tasked to recover air group crews who had diverted with damaged aircraft to ground fields. Then towards the end of the cruise, the Q-birds were used to fly in ahead of the strike force and jam radars.

The last day on station for Det 3 was 19 January 1951, by which time the Det had flown 598.2 hours (181.2 at night) and made 144 traps (29 at night). Squadron personnel then left for CONUS after transferring their aircraft to the VC-35 Det on Princeton.

The second Korean War Det, Det 41, was commanded by LCDR R.

Above left, AA riddled tail of VC-33 AD-3Q BuNo 122875 flown by ENS R.H. Rohr on 24 November 1950. The aircraft was hit over the Yalu River. (Jim Sullivan collection) Above, VC-33 AD-4N BuNo 126985 "U.S. Mule" taxis forward for a mission over Korea. Pilot was J.E. Hill, note impressive combat scoreboard on the fuselage side. (Ginter collection) Below, two views of VC-33 AD-4Q BuNo 124042 (SS/45) cutting the tail off VC-12 AD-3 BuNo 122805 as it goes over the side of the USS Leyte (CV-32) on 14 January 1952. The VC-12 pilot was unhurt. (National Archives)

Hoffmeister and operated aboard the USS Bon Homme Richard (CV/CVA-31) from 20 May 1952 to 8 January 1953. The BHR arrived on station on 23 June in time to join in on the strikes against the North Korean hydroelectric power plants. During the deployment, the Det lost AD-3Q BuNo 122863 to AAA fire over Pyongyang on 11 July 1952.

The third war Det, Det 44, was commanded by LT J.E. Gill and was aboard the USS Lake Champlain (CVA-39) from 26 April to 4 December 1953. The "Champ" com-menced combat operations on 13 June as flagship TF-77.

Meanwhile, in the Atlantic, the squadron reported on and off the big deck carriers (CVB-41, CVB-42 and CVB-43) and the Essex class carriers not involved in Korea, making numer-ous trips to the Mediterranean and the North Atlantic.

The squadron entered the jet age for about a year beginning in November 1952 when it received two F3D-2 SkyKnights. By April 1953 there were eight F3Ds on-hand which

Above, VC-33 AD-2 BuNo 122251 at NAS Atlantic City, NJ, on 17 July 1951. (Howard Levy) Below, VC-33 AD-4N traps aboard the USS Antietam (CVA-36) on 14 January 1953. (USN)

were evaluated as night attack air-craft. The type was rejected and their strength was reduced to four in November 1953 and by April 1954 all the SkyKnights were withdrawn.

A number of squadron Skyraiders

VC-33 NAS ATLANTIC CITY, NJ:

May 50	AD-4N (8), TBM-3E (10), TBM-3N (2)
Jun 50	AD-4N (8), AD-4Q (3), AD-3N (1), AD-2Q (3), AD-1Q (2), TBM-3N (2)
Jul 50	AD-4N (8), AD-4Q (3), AD-3N (2), AD-3Q (1), AD-2Q (3), AD-1Q (3), TBM-3N (3)
Aug 50	AD-4N (9), AD-4Q (4), AD-3N (4), AD-2Q (3), AD-1Q (4), TBM-3N (3)
Sep 50	AD-4N (5), AD-4Q (2), AD-3 (2), AD-3N (2), AD-3Q (8), AD-2Q (7), AD-1Q (7), TBM-3N (10), TBM-3Q (1), TBM-3E (1)
Oct 50	AD-4N (5), AD-4Q (2), AD-3 (2), AD-3N (1), AD-3Q (8), AD-2Q (7), AD-1Q (7)
Nov 50	AD-4N (10), AD-4Q (2), AD-3 (5), AD-3N (2), AD-3Q (8), AD-2Q (7), AD-1Q (7)
Dec 50	AD-4N (11), AD-4Q (2), AD-3 (7), AD-3N (4), AD-3Q (4), AD-2Q (8), AD-1Q (7), TBM-3E (2)
Jan 51	AD-4N (3), AD-3 (8), AD-3N (6), AD-3Q (4), AD-2Q (8), AD-1Q (7), TBM-3E (1)
Feb 51	AD-4Q (2), AD-3 (9), AD-3N (7), AD-3Q (6), AD-2Q (8), AD-1Q (7)
Mar 51	AD-4N (1), AD-3 (8), AD-3N (3), AD-3Q (8), AD-2Q (8), AD-1Q (7), TBM-3E (2)
Apr 51	AD-4N (3), AD-3N (6), AD-1Q (7)
May 51	AD-4N (3), AD-3 (8), AD-3N (3), AD-3Q (2), AD-2Q (5), AD-1Q (6)
Jun 51	AD-4N (3), AD-3 (8), AD-3N (7), AD-4Q (2), AD-3Q (2), AD-2Q (5), AD-1Q (6)
Jul 51	AD-4N (3), AD-3 (8), AD-3N (6), AD-4Q (2), AD-3Q (3), AD-2 (2), AD-2Q (5), AD-1Q (6)
Aug 51	AD-4N (3), AD-3 (11), AD-3N (4), AD-4Q (2), AD-3Q (6), AD-2 (3), AD-2Q (5), AD-1Q (6)
Sep 51	AD-4N (1), AD-3 (11), AD-3N (4), AD-4Q (2), AD-3Q (6), AD-2 (3), AD-2Q (5), AD-1Q (6)
Oct 51	AD-4NL (3), AD-3 (10), AD-3N (1), AD-4Q (3), AD-3Q (4), AD-2 (3), AD-2Q (7), AD-1Q (3)
Nov 51	AD-4 (3), AD-4NL (3), AD-3 (8), AD-3N (1), AD-4Q (2), AD-3Q (5), AD-2 (3), AD-2Q (6), AD-1Q (2)
Dec 51	AD-4 (3), AD-4NL (9), AD-4N (2), AD-3 (5), AD-3N (2), AD-4Q (3), AD-3Q (4), AD-2Q (6), AD-1Q (2)
Jan 52	AD-4 (3), AD-4NL (8), AD-4N (1), AD-3 (3), AD-3N (2), AD-4Q (1), AD-3Q (4), AD-2 (3), AD-2Q (5), AD-1Q (2)
Feb 52	AD-4 (3), AD-4NL (10), AD-4N (5), AD-3 (3), AD-3N (2), AD-3Q (3), AD-2 (3), AD-2Q (5), AD-1Q (2)
Mar 52	AD-4 (2), AD-4B (1), AD-4NL (10), AD-4N (6), AD-3 (3), AD-3N (2), AD-4Q (4), AD-2 (3), AD-2Q (6), AD-1 (2), AD-1Q (2)
Apr 52	AD-4 (2), AD-4B (1), AD-4NL (6), AD-4N (5), AD-3 (3), AD-3N (2), AD-4Q (4), AD-2 (3), AD-2Q (5), AD-1 (2), AD-1Q (3)
May 52	AD-4 (2), AD-4B (5), AD-4NL (2), AD-4N (4), AD-3 (3), AD-3N (2), AD-3Q (3), AD-2 (3), AD-2Q (4), AD-1 (2), AD-1Q (3)
Jun 52	AD-4 (2), AD-4B (9), AD-4NL (3), AD-4N (4), AD-3 (3), AD-3N (1), AD-3Q (3), AD-2 (3), AD-2Q (2), AD-1 (2), AD-1Q (3), TBM-3N (4)
Jul 52	AD-4 (2), AD-4B (5), AD-4NL (3), AD-4N (4), AD-3 (3), AD-3N (1), AD-4Q (1), AD-3Q (1), AD-2 (2), AD-2Q (1), AD-1 (2), AD-1Q (3), TBM-3N (4)
Aug 52	AD-4B (12), AD-4NL (5), AD-4N (11), AD-3 (3), AD-3N (2), AD-4Q (1), AD-3Q (2), AD-2 (2), AD-2Q (1), AD-1 (2), AD-1Q (2), TBM-3N (4)
Sep 52	AD-4B (5), AD-4NL (2), AD-4N (18), AD-3 (1), AD-4Q (1), AD-3Q (1), AD-2 (2), AD-2Q (1), AD-1 (2), TBM-3N (2)
Oct 52	AD-4B (7), AD-4NL (9), AD-4N (28), AD-3 (1), AD-3N (2), AD-4Q (3), AD-3Q (2), AD-2Q (1), TBM-3E (1)
Nov 52	AD-4B (7), AD-4NL (9), AD-4N (27), AD-3 (1), AD-3N (3), AD-4Q (3), AD-3Q (2), AD-2Q (1), F3D-2 (2)
Dec 52	AD-4B (8), AD-4NL (9), AD-4N (29), AD-3 (1), AD-3N (4), AD-4Q (2), AD-3Q (2), AD-2Q (1), F3D-2 (2)
Jan 53	AD-4B (8), AD-4NL (9), AD-4N (26), AD-3 (1), AD-3N (4), AD-4Q (1), AD-3Q (1), AD-2Q (1), F3D-2 (3)
Feb 53	AD-4B (8), AD-4NL (10), AD-4N (29), AD-3 (1), AD-3N (3), AD-4Q (1), F3D-2 (5)
Mar 53	AD-4B (12), AD-4NL (12), AD-4N (29), AD-3N (4), AD-4Q (1), AD-3Q (1), F3D-2 (8)
Apr 53	AD-4B (4), AD-4NL (9), AD-4N (21), AD-3N (4), AD-4Q (1), AD-3Q (1), F3D-2 (8)
May 53	AD-4B (8), AD-4NL (9), AD-4N (21), AD-3N (3), AD-3Q (1), F3D-2 (8)
Jun 53	AD-4B (11), AD-4NL (9), AD-4N (21), AD-3N (2), AD-3Q (1), F3D-2 (8)
Jul 53	AD-4B (11), AD-4NL (7), AD-4N (21), AD-3N (2), AD-4Q (1), AD-3Q (1), F3D-2 (8)
Aug 53	AD-4B (11), AD-4NL (6), AD-4N (13), AD-3N (2), AD-3Q (1), F3D-2 (8)
Sep 53	AD-4B (10), AD-4NL (8), AD-4N (13), AD-3N (2), AD-3Q (1), F3D-2 (8)
Oct 53	AD-4B (11), AD-4NL (8), AD-4N (18), AD-3N (2), AD-3Q (1), F3D-2 (8)
Nov 53	AD-4B (11), AD-4NL (3), AD-4N (17), AD-3N (2), AD-3Q (1), F3D-2 (4)
Dec 53	AD-4B (8), AD-4NL (3), AD-4N (22), AD-3N (2), AD-4Q (1), F3D-2 (4)
Jan 54	AD-4B (8), AD-4NL (9), AD-4N (26), AD-3N (4), AD-4Q (1), AD-3Q (1), AD-2Q (1), F3D-2 (3)
Feb 54	AD-4B (6), AD-4NL (6), AD-4N (16), AD-3N (2), AD-4Q (1), F3D-2 (4)
Mar 54	AD-4B (6), AD-4NL (6), AD-4N (20), AD-3N (1), AD-4Q (1), F3D-2 (1)
Apr 54	AD-5N (6), AD-4B (6), AD-4NL (1), AD-4N (18), AD-3N (2), AD-4Q (1)
May 54	AD-5N (11), AD-4B (6), AD-4NL (7), AD-4N (19), AD-3N (2), AD-3Q (1)
Jun 54	AD-5N (16), AD-4B (5), AD-4NL (7), AD-4N (19), AD-3N (2), AD-4Q (1), AD-3Q (2)
Jul 54	AD-5N (15), AD-4B (5), AD-4NL (7), AD-4N (17), AD-3N (2), AD-4Q (1), AD-3Q (1)
Aug 54	AD-5 (6), AD-5N (18), AD-4B (3), AD-4NL (7), AD-4N (16), AD-3N (1), AD-4Q (1), AD-3Q (2)
Sep 54	AD-5 (6), AD-5N (33), AD-4B (3), AD-4NL (5), AD-4N (13), AD-3N (1), AD-4Q (1)
Oct 54	AD-5 (6), AD-5N (34), AD-4NL (6), AD-4N (7), AD-3N (1), AD-4Q (4), AD-3Q (2)
Nov 54	AD-5 (6), AD-5N (29), AD-4NL (6), AD-4N (2), AD-3Q (2)
Dec 54	AD-5 (6), AD-5N (48), AD-4N (4), AD-4Q (1), AD-3Q (3)
Jan 55	AD-5 (6), AD-5N (44), AD-4Q (1), AD-3Q (3)
Feb 55	AD-5 (6), AD-5N (44), AD-4Q (1), AD-3Q (1)
Mar 55	AD-5 (6), AD-5N (44), AD-4Q (3), AD-3Q (1)
Apr 55	AD-5 (6), AD-5N (32), AD-4Q (2), AD-3Q (3)
May 55	AD-5 (4), AD-5N (29), AD-3Q (1)
Jun 55	AD-5 (4), AD-5N (33), AD-4Q (1), AD-3Q (3)
Jul 55	AD-5N (36), AD-4Q (3), AD-3Q (4)
Aug 55	AD-5N (33), AD-4Q (1), AD-3Q (4)
Sep 55	AD-5N (25), AD-4Q (1), AD-3Q (3)
Oct 55	AD-5N (23), AD-4Q (2), AD-3Q (4)
Nov 55	AD-5N (24), AD-4Q (1), AD-3Q (2)
Dec 55	AD-5N (28), AD-4Q (1), AD-3Q (1)
Jan 56	AD-5N (29)
Feb 56	AD-5N (30), AD-4Q (1)
Mar 56	AD-5N (25), AD-3Q (2)

Below, VC-33 AD-5N in flight with squadron's Night Hawks nickname on the engine cowl. (Paul Minert collection)

Apr 56   AD-5N (37), AD-3Q (1)
May 56   AD-5N (37)
Jun 56   AD-5N (40)
VA(AW)-33 NAS ATLANTIC CITY, NJ:
Jul 56   AD-5N (39)
Aug 56   AD-5N (39)
Sep 56   AD-5N (37)
Oct 56   AD-5N (36), S2F-2 (2)
Nov 56   AD-5N (38), S2F-2 (2)
Dec 56   AD-5N (39), S2F-2 (2)
Jan 57   AD-5N (29), S2F-2 (2)
Feb 57   AD-5N (31), S2F-2 (2), TF-1Q (2)
Mar 57   AD-5N (36), S2F-2 (1), S2F-1 (1), TF-1Q (2)
Apr 57   AD-5N (35), S2F-2 (1), S2F-1 (1), TF-1Q (2)
May 57   AD-5N (43), S2F-2 (1), S2F-1 (1), TF-1Q (2)
Jun 57   AD-5N (42), S2F-2 (2), TF-1Q (2)
Jul 57   AD-5N (45), S2F-2 (1), TF-1Q (2)
Aug 57   AD-5N (26), S2F-2 (1), TF-1Q (2)
Sep 57   AD-5N (21), S2F-2 (1), TF-1Q (2)
Oct 57   AD-5N (36), S2F-2 (1), TF-1Q (2)
Nov 57   AD-5N (31), TF-1Q (2)
Dec 57   AD-5N (32), S2F-2 (1), TF-1Q (2)
Jan 58   AD-5N (22), AD-5Q (6), S2F-2 (1), TF-1Q (2)
Feb 58   AD-5N (25), AD-5Q (8), TF-1Q (2)
Mar 58   AD-5N (26), AD-5Q (8), S2F-2 (1), TF-1Q (2)
VA(AW)-33 NAS QUONSET POINT, RI:
Apr 58   AD-5N (28), AD-5Q (10), S2F-1 (1), TF-1Q (2)
May 58   AD-5N (27), AD-5Q (10), S2F-2 (1), TF-1Q (2)
Jun 58   AD-5N (22), AD-5Q (11), S2F-2 (1), TF-1Q (2)
Jul 58   AD-5N (20), AD-5Q (11), S2F-2 (1), TF-1Q (2)
Aug 58   AD-5N (21), AD-5Q (12), S2F-2 (1), TF-1Q (2)
Sep 58   AD-5N (18), AD-5Q (13), S2F-2 (1), TF-1Q (2)
Oct 58   AD-5N (16), AD-5Q (14), S2F-2 (1), TF-1Q (2)
Nov 58   AD-5N (23), AD-5Q (14), S2F-2 (1), TF-1Q (2)
Dec 58   AD-5N (21), AD-5Q (15), S2F-2 (1), TF-1Q (2)
Jan 59   AD-5N (17), AD-5Q (13), TF-1Q (2)
Feb 59   AD-5N (13), AD-5Q (14), S2F-2 (1), TF-1Q (2)
Mar 59   AD-5N (8), AD-5Q (14), S2F-2 (1), TF-1Q (2)
Apr 59   AD-5N (8), AD-5Q (8), S2F-2 (1), TF-1Q (2)
May 59   AD-5N (8), AD-5Q (9), TF-1Q (2)
Jun 59   AD-5N (7), AD-5Q (11), TF-1Q (2)
VAW-33 NAS QUONSET POINT, RI:
Jul 59   AD-5N (6), AD-5Q (8), TF-1Q (2)
Aug 59   AD-5N (6), AD-5Q (9), TF-1Q (2)
Sep 59   AD-5N (6), AD-5Q (10), TF-1Q (2)
Oct 59   AD-5N (3), AD-5Q (14), TF-1Q (2)
Nov 59   AD-5Q (11), TF-1Q (2)
Dec 59   AD-5Q (8), TF-1Q (2)
Jan 60   AD-5Q (6), TF-1Q (2)
Feb 60   AD-5Q (7), TF-1Q (2)
Mar 60   AD-5Q (9), TF-1Q (2)
Apr 60   AD-5Q (6), TF-1Q (2)
May 60   AD-5Q (6), TF-1Q (2)
Jun 60   AD-5Q (6), TF-1Q (2)
Jul 60   AD-5Q (6), TF-1Q (2)
Aug 60   AD-5Q (8), TF-1Q (2)
Sep 60   AD-5Q (9), TF-1Q (2)
Oct 60   AD-5Q (9), TF-1Q (2)
Nov 60   AD-5Q (9), TF-1Q (2)
Dec 60   AD-5Q (9), TF-1Q (2)
Jan 61   AD-5Q (7), TF-1Q (2)
Feb 61   AD-5Q (2), TF-1Q (2)
Mar 61   AD-5Q (8), TF-1Q (2)
Apr 61   AD-5Q (8), TF-1Q (2)
May 61   AD-5Q (9), TF-1Q (2)
Jun 61   AD-5Q (6), TF-1Q (2)
Jul 61   AD-5Q (6), AD-5W (4), TF-1Q (2)
Aug 61   AD-5Q (8), AD-5W (18), TF-1Q (2)
Sep 61   AD-5Q (8), AD-5W (10), TF-1Q (1)
Oct 61   AD-5Q (8), AD-5W (14), TF-1Q (2)
Nov 61   AD-5Q (6), AD-5 (1), AD-5W (18), TF-1Q (2)
Dec 61   AD-5Q (9), AD-5W (12), TF-1Q (1)
Jan 62   AD-5Q (6), AD-5W (12), TF-1Q (1)
Feb 62   AD-5Q (7), AD-5 (1), AD-5W (14), TF-1Q (1)
Mar 62   AD-5Q (10), AD-5 (1), AD-5W (15), TF-1Q (1)

Apr 62   AD-5Q (2), AD-5 (1), AD-5W (10)
May 62   AD-5Q (12), AD-5 (1), AD-5W (11), TF-1Q (2)
Jun 62   AD-5Q (9), AD-5 (1), AD-5W (19), TF-1Q (2)
Jul 62   AD-5Q (6), AD-5W (13), TF-1Q (2)
Aug 62   AD-5Q (7), AD-5 (1), AD-5W (18), TF-1Q (1)
Sep 62   AD-5Q (7), AD-5 (1), AD-5W (16), TF-1Q (2)
Oct 62   EA-1F (6), EA-1E (16), EC-1A (2)
Nov 62   EA-1F (6), EA-1E (13), A-1E (1)
Dec 62   EA-1F (6), EA-1E (17), A-1E (1), EC-1A (1)
Jan 63   A-1H (2), EA-1F (2), EA-1E (17), EC-1A (1)
Feb 63   EA-1F (4), EA-1E (14), A-1E (1), EC-1A (1)
Mar 63   EA-1F (7), EA-1E (14), A-1E (1)
Apr 63   EA-1F (8), EA-1E (14), A-1E (1), EC-1A (1)
May 63   EA-1F (8), EA-1E (13), A-1E (1), EC-1A (1)
Jun 63   EA-1F (8), EA-1E (9), A-1E (1)
Jul 63   EA-1F (5), EA-1E (15), A-1E (1), EC-1A (2)
Aug 63   EA-1F (7), EA-1E (12), A-1E (1), EC-1A (2)
Sep 63   EA-1F (9), EA-1E (12), A-1E (1), EC-1A (2)
Oct 63   EA-1F (3), EA-1E (9), A-1E (1), EC-1A (2)
Nov 63   EA-1F (9), EA-1E (12), A-1E (1), EC-1A (2)
Dec 63   EA-1F (10), EA-1E (16), A-1E (1), EC-1A (2)
Jan 64   EA-1F (9), EA-1E (11), A-1E (1), EC-1A (2)
Feb 64   EA-1F (5), EA-1E (7), A-1E (1), UA-1E (1), EC-1A (2)
Mar 64   EA-1F (5), EA-1E (10), A-1E (1), UA-1E (1)
Apr 64   EA-1F (5), EA-1E (13), A-1E (1), UA-1E (1), EC-1A (2)
May 64   EA-1F (2), EA-1E (9), A-1E (1), EC-1A (2)
Jun 64   EA-1F (5), EA-1E (11), A-1E (1), EC-1A (1)
Jul 64   EA-1F (2), EA-1E (6), A-1E (1), UA-1E (1), EC-1A (1)
Aug 64   EA-1E (8), A-1E (1), UA-1E (1), EC-1A (2)
Sep 64   EA-1F (2), EA-1E (8), A-1E (1), UA-1E (1), EC-1A (2)
Oct 64   EA-1F (6), EA-1E (3), A-1E (1), UA-1E (1), EC-1A (1)
Nov 64   EA-1F (6), EA-1E (3), A-1E (1), UA-1E (1), EC-1A (1)
Dec 64   EA-1F (8), EA-1E (9), A-1E (1), UA-1E (1), EC-1A (2)
Jan 65   EA-1F (6), EA-1E (7), A-1E (1), UA-1E (1), EC-1A (2)
Feb 65   EA-1F (9), EA-1E (10), A-1E (1), UA-1E (1), EC-1A (2)
Mar 65   EA-1F (9), EA-1E (10), A-1E (1), UA-1E (1), EC-1A (2)
Apr 65   EA-1F (5), EA-1E (6), A-1E (1), EC-1A (2)
May 65   EA-1F (3), EA-1E (12), A-1E (1), EC-1A (2)
Jun 65   EA-1F (1), EA-1E (5), A-1E (1), EC-1A (2)
Jul 65   EA-1F (9), EA-1E (8), A-1E (1), EC-1A (2)
Aug 65   EA-1F (3), EA-1E (12), A-1E (1), EC-1A (2)
Sep 65   EA-1F (8), EA-1E (10), EC-1A (2)
Oct 65   EA-1F (1), EA-1E (8), EC-1A (2)
Nov 65   EA-1F (4), EA-1E (6), EC-1A (1)
Dec 65   EA-1F (8), EA-1E (8)
Jan 66   EA-1F (5), EA-1E (5), EC-1A (1)
Feb 66   EA-1F (9), EA-1G (2)
Mar 66   EA-1F (10), EA-1E (6)
Apr 66   EA-1F (10), EA-1E (6)
May 66   EA-1F (10), EA-1E (6)
Jun 66   EA-1F (7), EA-1E (5), EC-1A (1)
Jul 66   EA-1F (6), EA-1E (7)
Aug 66   EA-1F (6), EA-1E (7)
Sep 66   EA-1F (6), EA-1E (3), EC-1A (1)
Oct 66   EA-1F (6), EA-1E (3), EC-1A (1)
Nov 66   EA-1F (3), EA-1E (3), EC-1A (1)
Dec 66   EA-1F (7), EC-1A (2)
Jan 67   EA-1F (7), EC-1A (2)
Feb 67   EA-1F (4), EC-1A (2)
Mar 67   EA-1F (4), EC-1A (2)
Apr 67   EA-1F (1), EC-1A (2)
May 67   EA-1F (1), EC-1A (2)
Jun 67   EA-1F (1), EC-1A (2)
Jul 67   EA-1F (4), EC-1A (2)
Aug 67   EA-1F (4), EC-1A (2)
Sep 67   EA-1F (9), EC-1A (1)
Oct 67   EA-1F (9), EC-1A (1)
Nov 67   EA-1F (10), EC-1A (1)
Dec 67   EA-1F (7), A-1E (1), EC-1A (1)
Jan 68   EA-1F (9), A-1E (1), EC-1A (1)
VAQ-33 NAS QUONSET POINT, RI:
Feb 68   EA-1F (7), A-1E (1), EC-1A (2)
Mar 68   EA-1F (9), A-1E (1), EC-1A (2)
Apr 68   EA-1F (6), A-1E (1), EC-1A (2)
May 68   EA-1F (6), A-1E (1), EC-1A (2)

Jun 68	EA-1F (3), A-1E (1), EC-1A (2)
Jul 68	EA-1F (6), A-1E (1), EC-1A (2)
Aug 68	EA-1F (6), A-1E (1), EC-1A (2)
Sep 68	EA-1F (5), A-1E (1), EC-1A (2)
Oct 68	EA-1F (5), A-1E (1), EC-1A (2)
Nov 68	EA-1F (6), A-1E (1), EC-1A (2)
Dec 68	EA-1F (7), A-1E (1), EC-1A (2)
Jan 69	EA-1F (9), EC-1A (1)
Feb 69	EA-1F (14), EC-1A (1)
Mar 69	EA-1F (12), EA-1E (1), EC-1A (2)
Apr 69	EA-1F (10), A-1E (1), EC-1A (2)
May 69	EA-1F (9), A-1E (1), EC-1A (2)
Jun 69	EC-1A (2)
Jul 69	None
Aug 69	None
Sep 69	None

VAQ-33 NAS NORFOLK, VA:

Oct 69	F-4B (1), A-3B (1), EC-121K (1), TA-4F (4)
Nov 69	RA-3B (2)
Dec 69	EA-1F (1)
Jan 70	EA-1F (1), F-4B (1)

VC-33/VA(AW)-33/VAW-33/VAQ-33 DETS Sept. 1950 to Nov. 1969

Sep 50	Det 3	CV-32	AD-4N (2), AD-4Q (2)
	Det 6	CVB-43	AD-4N (4), AD-4Q (2)
Oct 50	Det 3	CV-32	AD-4N (2), AD-4Q (2)
	Det 6	CVB-43	AD-4N (4), AD-4Q (2)
	Det 7	CVB-42	AD-4N (3), AD-3N (1)
Nov 50	Det 3	CV-32	AD-4N (2), AD-4Q (2)
	Det 6	CVB-43	AD-4N (4), AD-4Q (2)
	Det 7	CVB-42	AD-4N (3), AD-3N (1)
Dec 50	Det 3	CV-32	AD-4N (2), AD-4Q (2)
	Det 6	CVB-43	AD-4N (4), AD-4Q (2)
	Det 13	CVL-48	AD-3N (2), AD-3Q (4)
Jan 51	Det 6	CVB-43	AD-4N (4), AD-4Q (2)
	Det 7	CVB-42	AD-4N (3), AD-3N (1)
	Det 11	CVL-49	AD-4N (4), AD-4Q (2)
Feb 51	Det 7	CVB-42	AD-4N (3), AD-3N (1)
	Det 8	CV-34	AD-3Q (4), TBM-3E (1)
	Det 11	CVL-49	AD-4N (4), AD-4Q (2)
	Det 26A	Argentia	AD-4N (6)
Mar 51	Det 6	CVB-43	AD-4N (4), AD-4Q (2)
	Det 7	CVB-42	AD-4N (3), AD-3N (1)
	Det 11	CVL-49	AD-4N (4), AD-4Q (2)
	Det 13	CVL-48	AD-3N (4), AD-4Q (2)
	Det 26A	Argentia	AD-4N (2)
Apr 51	Det 6	CVB-43	AD-4N (4), AD-4Q (2)
	Det	CV-34	AD-3Q (4)
	Det 13	CVL-48	AD-3N (4), AD-4Q (2)
May 51	Det 6	CVB-43	AD-4N (4), AD-4Q (2)
	Det 8	CV-34	AD-4N (2), AD-4Q (2)
	Det 13	CVL-48	AD-3N (4), AD-4Q (2)
Jun 51	Det 6	CVB-43	AD-4N (4), AD-4Q (2)
	Det 8	CV-34	AD-4N (2), AD-4L (1)
Jul 51	Det 3	CV-32	AD-3Q (3)
	Det 6	CVB-43	AD-4N (4), AD-4Q (2)
	Det 8	CV-34	AD-4N (2), AD-4L (1), AD-4Q (2)
Aug 51	Det 6	CVB-43	AD-4N (4), AD-4Q (2)
	Det 8	CV-34	AD-4N (2), AD-4L (1), AD-4Q (2)
Sep 51	Det 3	CV-32	AD-4N (2), AD-4Q (1)
	Det 6	CVB-43	AD-4N (4), AD-4Q (2)
	Det 7	CVB-42	AD-4Q (1)
	Det 8	CV-34	AD-4N (2), AD-4L (1), AD-4Q (2)
Oct 51	Det 3	CV-32	AD-4NL (1), AD-4N (2), AD-4Q (1)
	Det 5	CVB-41	AD-4N (1), AD-3N (3)
	Det 7	CVB-42	AD-4N (4), AD-4Q (1)
Nov 51	Det 3	CV-32	AD-4NL (1), AD-4N (2), AD-4Q (1)
	Det 7	CVB-42	AD-4N (4), AD-4Q (1)
	Det 32	CV-40	AD-4N (1), AD-3N (2), AD-4Q (2)
Dec 51	Det 7	CVB-42	AD-4N (4), AD-4Q (1)
	Det 32	CV-40	AD-4NL (1), AD-4N (1), AD-3N (2), AD-4Q (1)
Jan 52	Det 5	CVB-41	AD-4NL (4), AD-4Q (1)
	Det 7	CVB-42	AD-4N (4), AD-4Q (1)
	Det 32	CV-40	AD-4NL (1), AD-4N (1), AD-3N (2), AD-4Q (1)
Feb 52	Det 5	CVB-41	AD-4NL (4), AD-4Q (1)
	Det 32	CV-40	AD-4NL (1), AD-4N (1), AD-3N (2), AD-4Q (1)
Mar 52	Det 5	CVB-41	AD-4NL (4), AD-4Q (1)
	Det 32	CV-40	AD-4NL (1), AD-4N (1), AD-3N (2), AD-4Q (1)
Apr 52	Det 5	CVB-41	AD-4Q (1)
	Det 6	CVB-43	AD-4B (4), AD-4NL (4), AD-4Q (1), TBM-3E (1)
	Det 32	CV-40	AD-4NL (1), AD-4N (1), AD-3W (2)
	Det 38	CV-18	AD-4NL (4), TBM-3E (1)
May 52	Det 6	CVB-43	AD-4B (4), AD-4NL (4), AD-4Q (1), TBM-3E (1)
	Det 32	CV-40	AD-4NL (1), AD-4N (1), AD-3N (2), AD-4Q (1)
	Det 38	CV-18	AD-4NL (4), AD-4Q (1)
Jun 52	Det 5	CVB-41	AD-3N (3), AD-4Q (1)
	Det 6	CVB-43	AD-4B (4), AD-4NL (4), TBM-3E (1)
	Det 38	CV-18	AD-4NL (4), AD-4Q (1)
Jul 52	Det 5	CVB-41	AD-3N (3), AD-4Q (1)
	Det 6	CVB-43	AD-4B (4), AD-4NL (4), TBM-3E (1)
	Det 7	CVB-42	AD-4B (4), AD-4N (1), AD-4Q (1), AD-3Q (2), AD-2Q (1)
	Det 38	CV-18	AD-4NL (4), AD-4Q (1)
Aug 52	Det 3	CV-32	AD-4N (4), AD-4Q (1)
	Det 5	CVB-41	AD-4N (4), AD-4Q (1) AD-3Q (1)
	Det 6	CVB-43	AD-4B (3), AD-4NL (4), TBM-3E (1)
	Det 7	CVB-42	AD-4B (4), AD-4N (4)
	Det 38	CV-18	AD-4NL (2), AD-4Q (1)
Sep 52	Det 3	CV-32	AD-4NL (1), AD-4N (4), AD-4Q (1)
	Det 5	CVB-41	AD-4N (4), AD-4Q (1) AD-3Q (1)
	Det 6	CVB-43	AD-4B (3), AD-4NL (4), TBM-3E (1)
	Det 7	CVB-42	AD-4B (4), AD-4N (4)
	Det 38	CV-18	AD-4NL (2), AD-4Q (1)
Oct 52	Det 3	CVA-32	AD-4N (4), AD-4Q (1)
	Det 5	CVA-41	AD-3Q (1)
	Det 7	CVA-42	AD-4B (5), AD-4N (4)
Nov 52	Det 3	CVA-32	AD-4N (3), AD-4Q (1)
	Det 5	CVA-41	AD-3Q (1)
	Det 7	CVA-42	AD-4B (4), AD-4N (3)
Dec 52	Det 3	CVA-32	AD-4N (3), AD-4Q (1)
	Det 5	CVA-41	AD-4B (4), AD-4N (4), AD-4Q (1), AD-3Q (1)
Jan 53	Det 3	CVA-32	AD-4N (3)
	Det 5	CVA-41	AD-4B (4), AD-4N (4), AD-4Q (2), AD-3Q (1)
	Det 32	CVA-40	AD-4N (4), AD-4Q (1)
Feb 53	Det 5	CVA-41	AD-4B (4), AD-4N (4), AD-4Q (1)
	Det 32	CVA-40	AD-4N (4), AD-4Q (1)
Mar 53	Det 5	CVA-41	AD-4B (4), AD-4N (4), AD-4Q (1)
	Det 32	CVA-40	AD-4N (4), AD-4Q (1)
Apr 53	Det 5	CVA-41	AD-4B (4), AD-4N (4), AD-4Q (1)
	Det 6	CVA-43	AD-4B (4), AD-4N (4), AD-3Q (1)
	Det 32	CVA-40	AD-4N (4), AD-4Q (1)
	Det 44	CVA-39	AD-4N (4)
May 53	Det 6	CVA-43	AD-4B (4), AD-4N (4), AD-3Q (1)
	Det 7	CVA-42	AD-4N (4), AD-4Q (1)
	Det 32	CVA-40	AD-4N (4), AD-4Q (1)
	Det 44	CVA-39	AD-4N (4)
Jun 53	Det 6	CVA-43	AD-4B (1), AD-4N (4), AD-3Q (1)
	Det 7	CVA-42	AD-4N (4), AD-4Q (1)
	Det 32	CVA-40	AD-4N (4), AD-4Q (1)
	Det 44	CVA-39	AD-4N (3)
Jul 53	Det 6	CVA-43	AD-4B (1), AD-4N (4), AD-3Q (1)
	Det 7	CVA-42	AD-4N (4), AD-4Q (1)
	Det 41	Quonset	AD-4N (4)
	Det 44	CVA-39	AD-4N (4)
Aug 53	Det 6	CVA-43	AD-4B (2), AD-4N (4), AD-4 (1)AD-3Q (1)
	Det 7	CVA-42	AD-4N (4), AD-4Q (1)
	Det 41	Quonset	AD-4N (6), AD-4Q (1)
	Det 44	CVA-39	AD-4N (4)
	Det 53	CVA-45	AD-4N (4)
Sep 53	Det 6	CVA-43	AD-4B (2), AD-4N (4), AD-4 (1)AD-3Q (1)
	Det 7	CVA-42	AD-4N (4), AD-4Q (1)
	Det 38	CVA-18	AD-4N (4)
	Det 41	Quonset	AD-4N (6), AD-4Q (1)
	Det 44	CVA-39	AD-4N (4)
Oct 53	Det 7	CVA-42	AD-4N (4), AD-4Q (1)
	Det 32	CVA-40	AD-4N (4)
	Det 38	CVA-18	AD-4N (4)
	Det 41	Quonset	AD-4N (6), AD-4Q (1)
	Det 44	CVA-39	AD-4N (3)
Nov 53	Det 5	CVA-41	AD-4NL (4), AD-3Q (1)
	Det 7	CVA-42	AD-4N (4), AD-4Q (1)
	Det 32	CVA-40	AD-4N (4)
	Det 38	CVA-18	AD-4N (2)
	Det 41	Quonset	AD-4N (5), AD-4Q (1)
	Det 44	CVA-39	AD-4N (3)
Dec 53	Det 5	CVA-41	AD-4NL (4), AD-3Q (1)
	Det 7	CVA-42	AD-4N (1)
	Det 32	CVA-40	AD-4N (3)
	Det 38	CVA-18	AD-4N (2)
	Det 41	Quonset	AD-4N (5), AD-4Q (1)
	Det 44	CVA-39	AD-4N (3)
Jan 54	Det 3	CVA-32	AD-4N (3)
	Det 5	CVA-41	AD-4B (4), AD-4N (4), AD-4Q (2), AD-3Q (1)
	Det 32	CVA-40	AD-4N (4), AD-4Q (1)
Feb 54	Det 3	CVA-32	AD-4N (3)
	Det 32	CVA-40	AD-4N (4)
	Det 38	CVA-18	AD-4N (4)
	Det 5	CVA-41	AD-4NL (4), AD-3Q (1)
	Det 57	CVA-12	AD-4N (4)
Mar 54	Det 35	CVA-41	AD-4NL (4), AD-3Q (1)
	Det 38	CVA-40	AD-4N (4)
	Det 40	CVA-18	AD-4N (4)
Apr 54	Det 31	CVA-43	AD-4NL (4)
	Det 35	CVA-41	AD-4NL (4), AD-3Q (1)
	Det 38	CVA-40	AD-4N (4)
	Det 40	CVA-18	AD-4N (4)
May 54	Det 30	CVA-20	AD-4N (4), AD-4Q (1)
	Det 31	CVA-43	AD-4NL (4)
	Det 35	CVA-41	AD-4NL (3), AD-3Q (1)
	Det 38	CVA-40	AD-4N (4)
Jun 54	Det 31	CVA-43	AD-4NL (4)
	Det 35	CVA-41	AD-4NL (2), AD-3Q (1)
	Det 38	CVA-40	AD-4N (4)
	Det 52	CVS-45	AD-4N (2)
Jul 54	Det 31	CVA-43	AD-4N (4), AD-4Q (1), AD-3Q (1)
	Det 35	CVA-41	AD-4NL (2), AD-3Q (1)
	Det 38	CVA-40	AD-4N (4)
Aug 54	Det 31	CVA-43	AD-4N (4), AD-4Q (1), AD-3Q (1)

Date	Det	Ship	Aircraft
	Det 34	CVA-39	AD-4NL (2), AD-4N (2), AD-4Q (1), AD-3Q (1)
Sep 54	Det 31	CVA-43	AD-4N (4), AD-4Q (1), AD-3Q (1)
	Det 34	CVA-39	AD-4NL (2), AD-4N (2), AD-4Q (1), AD-3Q (1)
Oct 54	Det 31	CVA-43	AD-4N (4), AD-4Q (1), AD-3Q (1)
	Det 34	CVA-39	AD-4NL (2), AD-4N (2), AD-4Q (1), AD-3Q (1)
Nov 54	Det 31	CVA-43	AD-4N (4), AD-4Q (1), AD-3Q (1)
	Det 34	CVA-39	AD-4NL (2), AD-4N (3), AD-4Q (1), AD-3Q (1)
	Det 36	CVA-15	AD-5N (5), AD-4Q (1), AD-3Q (1)
Dec 54	Det 34	CVA-39	AD-4NL (2), AD-4N (3), AD-4Q (1), AD-3Q (1)
	Det 36	CVA-15	AD-5N (5), AD-4Q (1), AD-3Q (1)
Jan 55	Det 34	CVA-39	AD-4NL (1), AD-4N (3), AD-4Q (1), AD-3Q (1)
	Det 35	CVA-41	AD-5N (4)
	Det 36	CVA-15	AD-5N (4), AD-4Q (1), AD-3Q (1)
Feb 55	Det 34	CVA-39	AD-4NL (1), AD-4N (3), AD-3Q (1)
	Det 35	CVA-41	AD-5N (4)
	Det 36	CVA-15	AD-5N (5), AD-4Q (1), AD-3Q (1)
Mar 55	Det 31	CVA-43	AD-5N (5), AD-3Q (2)
	Det 34	CVA-39	AD-4NL (1), AD-4N (2), AD-3Q (1)
	Det 35	CVA-41	AD-5N (4)
	Det 36	CVA-15	AD-5N (4), AD-4Q (1), AD-3Q (1)
Apr 55	Det 31	CVA-43	AD-5N (4), AD-3Q (2)
	Det 32	Miramar	AD-5N (4)
	Det 33	CVA-11	AD-5N (4), AD-2Q (2)
	Det 35	CVA-41	AD-5N (4)
	Det 36	CVA-15	AD-5N (4), AD-4Q (1), AD-3Q (1)
May 55	Det 31	CVA-43	AD-5N (4), AD-3Q (2)
	Det 32	CVA-12	AD-5N (4)
	Det 33	CVA-11	AD-5N (4), AD-2Q (2)
	Det 35	CVA-41	AD-5N (4)
	Det 36	CVA-15	AD-5N (4), AD-4Q (1), AD-3Q (1)
	Det 58	CVE-112	AD-4Q (2), AD-3Q (2)
Jun 55	Det 31	CVA-43	AD-5N (4), AD-3Q (2), F2H-2P (1)
	Det 32	CVA-12	AD-5N (4)
	Det 33	CVA-11	AD-4W (3)
	Det 35	CVA-41	AD-5N (4)
	Det 58	CVE-112	AD-4Q (2), AD-3Q (1)
Jul 55	Det 31	CVA-43	AD-5N (4), AD-3Q (2)
	Det 32	CVA-12	AD-5N (4)
	Det 33	CVA-11	AD-5N (4), AD-4Q (2)
Aug 55	Det 31	CVA-43	AD-5N (4), AD-3Q (1)
	Det 32	CVA-12	AD-5N (4)
	Det 33	CVA-11	AD-5N (4), AD-4Q (2)
	Det 34	CVA-39	AD-5N (4), AD-4Q (2)
Sep 55	Det 31	CVA-43	AD-5N (4), AD-3Q (1)
	Det 33	CVA-11	AD-5N (4), AD-4Q (2)
	Det 34	CVA-39	AD-5N (4), AD-4Q (1)
	Det 52	CVS-45	AD-4Q (1), AD-3Q (1)
Oct 55	Det 30	CVA-20	AD-5N (4)
	Det 32	CVA-12	AD-5N (4)
	Det 33	CVA-11	AD-5N (4), AD-4Q (2), AJ-2 (1)
	Det 34	CVA-39	AD-5N (5), AD-4Q (1)
	Det 39	CVA-14	AD-5N (4), AD-3Q (2)
Nov 55	Det 30	CVA-20	AD-5N (4)
	Det 32	CVA-12	AD-5N (4)
	Det 34	CVA-39	AD-5N (4), AD-3Q (2)
	Det 39	CVA-14	AD-5N (3), AD-4Q (1)
Dec 55	Det 30	CVA-20	AD-5N (4)
	Det 34	CVA-39	AD-5N (4), AD-3Q (2)
	Det 39	CVA-14	AD-5N (3), AD-4Q (1)
Jan 56	Det 30	CVA-20	AD-5N (3)
	Det 34	CVA-39	AD-5N (4)
	Det 39	CVA-14	AD-5N (3), AD-4Q (1)
Feb 56	Det 30	CVA-20	AD-5N (4)
	Det 33	CVA-11	AD-5N (4)
	Det 34	CVA-39	AD-5N (4), AD-3Q (2)
	Det 39	CVA-14	AD-5N (3), AD-4Q (1)
Mar 56	Det 30	CVA-20	AD-5N (4)
	Det 33	CVA-11	AD-5N (4)
	Det 36	CVA-15	AD-5N (4)
	Det 39	CVA-14	AD-5N (5), AD-4Q (1)
	Det 42	CVA-59	AD-5N (4)
Apr 56	Det 33	CVA-11	AD-5N (4)
	Det 36	CVA-15	AD-5N (3)
	Det 39	CVA-14	AD-5N (4), AD-4Q (1)
May 56	Det 33	CVA-11	AD-5N (5)
Jun 56	Det 33	CVA-11	AD-5N (6)
	Det 39	CVA-14	AD-5N (4)
VA(AW)-33:			
Jul 56	Det 33	CVA-11	AD-5N (6)
	Det 36	CVA-15	AD-5N (4)
	Det 39	CVA-14	AD-5N (4), AD-4Q (1)
Aug 56	Det 33	CVA-11	AD-5N (5)
	Det 36	CVA-15	AD-5N (4)
	Det 39	CVA-14	AD-5N (3)
Sep 56	Det 31	CVA-43	AD-5N (1)
	Det 33	CVA-11	AD-5N (4)
	Det 36	CVA-15	AD-5N (4)
	Det 39	CVA-14	AD-5N (1)
Oct 56	Det 30	CVA-20	AD-5N (4)
	Det 31	CVA-43	AD-5N (1)
	Det 33	CVA-11	AD-5N (5)
	Det 36	CVA-15	AD-5N (4)
	Det 39	CVA-14	AD-5N (1)
Nov 56	Det 30	CVA-20	AD-5N (4)
	Det 31	CVA-43	AD-5N (4)
	Det 36	CVA-15	AD-5N (4)
	Det 42	CVA-59	AD-5N (3)
Dec 56	Det 30	CVA-20	AD-5N (4)
	Det 31	CVA-43	AD-5N (4)
	Det 36	CVA-15	AD-5N (4)
Jan 57	Det 31	CVA-43	AD-5N (3)
	Det 34	CVA-39	AD-5N (4)
	Det 36	CVA-15	AD-5N (4)
	Det 42	CVA-59	AD-5N (4)
Feb 57	Det 30	CVA-20	AD-5N (4)
	Det 31	CVA-43	AD-5N (3)
	Det 34	CVA-39	AD-5N (4)
	Det 36	CVA-15	AD-5N (4)
	Det 42	CVA-59	AD-5N (4)
Mar 57	Det 34	CVA-39	AD-5N (4)
	Det 36	CVA-15	AD-5N (1)
	Det 42	CVA-59	AD-5N (4)
Apr 57	Det 34	CVA-39	AD-5N (5)
	Det 36	CVA-15	AD-5N (3)
	Det 42	CVA-59	AD-5N (4)
May 57	Det 34	CVA-39	AD-5N (5)
	Det 36	CVA-15	AD-5N (3)
	Det 42	CVA-59	AD-5N (4)
Jun 57	Det 34	CVA-39	AD-5N (4)
	Det 36	CVA-15	AD-5N (3)
	Det 42	CVA-59	AD-5N (4)
Jul 57	Det 33	CVA-11	AD-5N (4)
	Det 34	CVA-39	AD-5N (5)
	Det 36	CVA-15	AD-5N (3)
	Det 42	CVA-59	AD-5N (4)
Aug 57	Det 33	CVA-11	AD-5N (4)
	Det 34	CVA-39	AD-5N (5)
	Det 36	CVA-15	AD-5N (3)
	Det 42	CVA-59	AD-5N (4)
	Det 43	CVA-60	AD-5N (6)
Sep 57	Det 33	CVA-11	AD-5N (4)
	Det 34	CVA-39	AD-5N (5)
	Det 36	CVA-15	AD-5N (3)
	Det 42	CVA-59	AD-5N (7)
	Det 43	CVA-60	AD-5N (7)
Oct 57	Det 36	CVA-15	AD-5N (3)
Nov 57	Det 36	CVA-15	AD-5N (4)
	Det 45	CVA-9	AD-5N (3)
Dec 57	Det 36	CVA-15	AD-5N (4)
	Det 45	CVA-9	AD-5N (1)
Jan 58	Det 36	CVA-15	AD-5N (4)
	Det 43	CVA-60	AD-5N (5)
	Det 45	CVA-9	AD-5N (6)
Feb 58	Det 36	CVA-15	AD-5N (1)
	Det 43	CVA-60	AD-5N (5)
	Det 45	CVA-9	AD-5N (4)
Mar 58	Det 43	CVA-60	AD-5N (5)
	Det 45	CVA-9	AD-5N (4)
Apr 58	Det 43	CVA-60	AD-5N (5)
	Det 45	CVA-9	AD-5N (4)
May 58	Det 33	CVA-11	AD-5N (3)
	Det 43	CVA-60	AD-5N (4)
	Det 45	CVA-9	AD-5N (4)
Jun 58	Det 33	CVA-11	AD-5N (3)
	Det 36	CVA-15	AD-5N (3)
	Det 43	CVA-60	AD-5N (4)
	Det 45	CVA-9	AD-5N (4)
Jul 58	Det 33	CVA-11	AD-5N (3)
	Det 36	CVA-15	AD-5N (3)
	Det 42	CVA-59	AD-5N (3)
	Det 43	CVA-60	AD-5N (3)
	Det 45	CVA-9	AD-5N (4)
Aug 58	Det 36	CVA-15	AD-5N (3)
	Det 42	CVA-59	AD-5N (3)
	Det 43	CVA-60	AD-5N (3)
	Det 45	CVA-9	AD-5N (4)
Sep 58	Det 36	CVA-15	AD-5N (3)
	Det 42	CVA-59	AD-5N (3)
	Det 43	CVA-60	AD-5N (2)
	Det 45	CVA-9	AD-5N (4)
Oct 58	Det 33	CVA-11	AD-5N (3)
	Det 36	CVA-15	AD-5N (3)
	Det 42	CVA-59	AD-5N (3)
	Det 45	CVA-9	AD-5N (4)
Nov 58	Det 36	CVA-15	AD-5N (3)
	Det 42	CVA-59	AD-5N (3)
Dec 58	Det 36	CVA-15	AD-5N (2)
	Det 42	CVA-59	AD-5N (3)
Jan 59	Det 33	CVA-11	AD-5Q (3)
	Det 36	CVA-15	AD-5N (2)
	Det 42	CVA-59	AD-5N (3)
Feb 59	Det 33	CVA-11	AD-5Q (3)
	Det 36	CVA-15	AD-5N (2)
	Det 42	CVA-59	AD-5N (3)
Mar 59	Det 33	CVA-11	AD-5Q (3)
	Det 36	CVS-15	AD-5N (1)
	Det 42	CVA-59	AD-5N (3)
	Det 43	CVA-60	AD-5N (1)
Apr 59	Det 7	Gitmo	AD-5Q (3)
	Det 33	CVA-11	AD-5Q (3)
	Det 36	CVS-15	AD-5N (1)
	Det 43	CVA-60	AD-5N (1)
May 59	Det 7	Gitmo	AD-5Q (3)
	Det 33	CVA-11	AD-5Q (3)
	Det 43	CVA-60	AD-5N (1)
Jun 59	Det 33	CVA-11	AD-5Q (2)
	Det 43	CVA-60	AD-5N (1), AD-5Q (4)
VAW-33:			
Jul 59	Det 33	CVA-11	AD-5Q (2)
	Det 43	CVA-60	AD-5N (1), AD-5Q (4)
Aug 59	Det 33	CVA-11	AD-5Q (2)
	Det 42	CVA-59	AD-5Q (3)
	Det 43	CVA-60	AD-5N (1), AD-5Q (3)
Sep 59	Det 42	CVA-59	AD-5Q (4)
	Det 43	CVA-60	AD-5N (1), AD-5Q (4)

```
Oct 59  Det 43  CVA-60   AD-5Q (3)
Nov 59  Det 43  CVA-60   AD-5Q (3)
Dec 59  Det 42  CVA-59   AD-5Q (3)
        Det 43  CVA-60   AD-5Q (3)
Jan 60  Det 42  CVA-59   AD-5Q (3)
        Det 43  CVA-60   AD-5Q (3)
Feb 60  Det 42  CVA-59   AD-5Q (3)
        Det 43  CVA-60   AD-5Q (2)
Mar 60  Det 42  CVA-59   AD-5Q (4)
Apr 60  Det 41  CVA-62   AD-5Q (3)
        Det 42  CVA-59   AD-5Q (4)
May 60  Det 41  CVA-62   AD-5Q (3)
        Det 42  CVA-59   AD-5Q (4)
Jun 60  Det 41  CVA-62   AD-5Q (3)
        Det 42  CVA-59   AD-5Q (4)
Jul 60  Det 41  CVA-62   AD-5Q (3)
        Det 42  CVA-59   AD-5Q (3)
Aug 60  Det 41  CVA-62   AD-5Q (3)
        Det 42  CVA-59   AD-5Q (2)
Sep 60  Det 41  CVA-62   AD-5Q (3)
Oct 60  Det 41  CVA-62   AD-5Q (3)
Nov 60  Det 41  CVA-62   AD-5Q (3)
Dec 60  Det 41  CVA-62   AD-5Q (3)
Jan 61  Det 41  CVA-62   AD-5Q (3)
Feb 61  Det 41  CVA-62   AD-5Q (3)
Mar 61  Det 42  CVA-59   AD-5Q (3)
Apr 61  Det 42  CVA-59   AD-5Q (3)
May 61  Det 42  CVA-59   AD-5Q (3)
Jun 61  Det 41  CVA-62   AD-5Q (3)
        Det 42  CVA-59   AD-5Q (3)
Jul 61  Det 34  CVS-39   AD-5W (4)
        Det 41  CVA-62   AD-5Q (2)
        Det 42  Sig. Italy AD-5Q (3)
Aug 61  Det 34  CVS-39   AD-5W (4)
        Det 41  CVA-62   AD-5Q (3)
        Det 42  Sig. Italy AD-5Q (3)
Sep 61  Det 34  CVS-39   AD-5W (4)
        Det 41  CVA-62   AD-5Q (3)
        Det 42  Sig. Italy AD-5Q (3)
        Det 45  CVS-9    AD-5W (4)
Oct 61  Det 41  CVA-62   AD-5Q (3)
        Det 42  Sig. Italy AD-5Q (3)
        Det 45  CVS-9    AD-5W (4)
Nov 61  Det 41  CVA-62   AD-5Q (3)
        Det 42  Sig. Italy AD-5Q (3)
        Det 45  CVS-9    AD-5W (4)
        Det 48  CVS-18   AD-5W (4)
Dec 61  Det 34  CVS-39   AD-5W (4)
        Det 42  Sig. Italy AD-5Q (3)
        Det 43  CVA-60   AD-5Q (3)
        Det 45  CVS-9    AD-5W (4)
Jan 62  Det 42  Sig. Italy AD-5Q (3)
        Det 43  CVA-60   AD-5Q (3)
        Det 45  CVS-9    AD-5W (4)
        Det 48  CVS-18   AD-5W (4)
Feb 62  Det 42  Sig. Italy AD-5Q (3)
        Det 43  CVA-60   AD-5Q (3)
        Det 48  CVS-18   AD-5W (4)
Mar 62  Det 42  Sig. Italy AD-5Q (3)
        Det 43  CVA-60   AD-5Q (3)
        Det 48  CVS-18   AD-5W (4)
Apr 62  Det 33  CVS-11   AD-5W (4)
        Det 34  CVS-39   AD-5W (4)
        Det 41  CVA-62   AD-5Q (3)
        Det 42  Sig. Italy AD-5Q (3)
        Det 43  CVA-60   AD-5Q (3)
        Det 48  CVS-18   AD-5W (4)
May 62  Det 11  CVS-11   AD-5W (4)
        Det 39  CVS-39   AD-5W (4)
        Det 62  CVA-62   AD-5Q (3)
        Det 18  CVS-18   AD-5W (4)
Jun 62  Det 11  CVS-11   AD-5W (4)
        Det 39  CVS-39   AD-5W (4)
        Det 59  CVA-59   AD-5Q (3)
        Det 62  CVA-62   AD-5Q (2)

Jul 62  Det 11  CVS-11   AD-5W (4)
        Det 39  CVS-39   AD-5W (4)
        Det 59  CVA-59   AD-5Q (3)
        Det 62  CVA-62   AD-5Q (3)
Aug 62  Det 39  CVS-39   AD-5W (4)
        Det 59  CVA-59   AD-5Q (3)
Sep 62  Det 39  CVS-39   AD-5W (5)
        Det 59  CVA-59   AD-5Q (3)
Oct 62  Det 18  CVS-18   EA-1E (4)
        Det 59  CVA-59   EA-1F (3)
        Det 62  CVA-62   EA-1F (3)
Nov 62  Det 39  CVS-39   EA-1E (5)
        Det 59  CVA-59   EA-1F (3)
        Det 60  CVA-60   EA-1F (3)
Dec 62  Det 59  CVA-59   EA-1F (3)
        Det 60  CVA-60   EA-1F (3)
Jan 63  Det 11  CVS-11   EA-1E (4)
        Det 18  CVS-18   EA-1E (4)
        Det 59  CVA-59   EA-1F (3)
        Det 60  CVA-60   EA-1F (3)
        Det 65  CVAN-65  EA-1F (3)
Feb 63  Det 11  CVS-11   EA-1E (4)
        Det 18  CVS-18   EA-1E (4)
        Det 59  CVA-59   EA-1F (3)
        Det 65  CVAN-65  EA-1F (3)
Mar 63  Det 18  CVS-18   EA-1E (4)
        Det 65  CVAN-65  EA-1F (3)
Apr 63  Det 18  CVS-18   EA-1E (4)
        Det 65  CVAN-65  EA-1F (3)
May 63  Det 18  CVS-18   EA-1E (4)
        Det 65  CVAN-65  EA-1F (3)
Jun 63  Det 11  CVS-11   EA-1E (4)
        Det 18  CVS-18   EA-1E (4)
        Det 65  CVAN-65  EA-1F (3)
Jul 63  Det 11  CVS-11   EA-1E (4)
        Det 62  CVA-62   EA-1F (3)
        Det 65  CVAN-65  EA-1F (3)
Aug 63  Det 11  CVS-11   EA-1E (4)
        Det 62  CVA-62   EA-1F (3)
        Det 65  CVAN-65  EA-1F (2)
Sep 63  Det 11  CVS-11   EA-1E (4)
        Det 39  CVS-39   EA-1F (4)
        Det 62  CVA-62   EA-1F (3)
        Det 65  CVAN-65  EA-1F (3)
Oct 63  Det 11  CVS-11   EA-1E (4)
        Det 39  CVS-39   EA-1F (4)
        Det 59  CVA-59   EA-1F (3)
        Det 62  CVA-62   EA-1F (3)
        Det 65  CVAN-65  EA-1F (3)
Nov 63  Det 11  CVS-11   EA-1E (4)
        Det 62  CVA-62   EA-1F (2)
Dec 63  Det 11  CVS-11   EA-1E (4)
        Det 62  CVA-62   EA-1F (3)
Jan 64  Det 11  CVS-11   EA-1E (4)
        Det 62  CVA-62   EA-1F (5)
Feb 64  Det 11  CVS-11   EA-1E (4)
        Det 39  CVS-39   EA-1E (4)
Mar 64  Det 39  CVS-39   EA-1E (5)
        Det 60  CVA-60   EA-1F (1)
        Det 65  CVAN-65  EA-1F (3)
Apr 64  Det 18  CVS-18   EA-1E (2)
        Det 59  CVA-59   EA-1F (3)
        Det 65  CVAN-65  EA-1F (3)
May 64  Det 11  CVS-11   EA-1E (2)
        Det 18  CVS-18   EA-1E (2)
        Det 39  CVS-39   UA-1F (1), EA-1E (3)
        Det 59  CVA-59   EA-1F (3)
        Det 60  CVA-60   EA-1F (3)
        Det 65  CVAN-65  EA-1F (3)
Jun 64  Det 11  CVS-11   EA-1E (4)
        Det 18  CVS-18   EA-1E (4)
        Det 59  CVA-59   EA-1F (3)
        Det 62  CVA-62   EA-1F (2)
        Det 65  CVAN-65  EA-1F (3)
Jul 64  Det 11  CVS-11   EA-1E (3)

        Det 18  CVS-18   EA-1E (4)
        Det 59  CVA-59   EA-1F (3)
        Det 62  CVA-62   EA-1F (2)
        Det 65  CVAN-65  EA-1F (3)
Aug 64  Det 11  CVS-11   EA-1E (4)
        Det 59  CVA-59   EA-1F (3)
        Det 62  CVA-62   EA-1F (4)
        Det 65  CVAN-65  EA-1F (3)
Sep 64  Det 18  CVS-18   EA-1E (4)
        Det 59  CVA-59   EA-1F (3)
        Det 62  CVA-62   EA-1F (4)
Oct 64  Det 39  CVS-39   EA-1E (2)
        Det 18  CVS-18   EA-1E (4)
        Det 59  CVA-59   EA-1F (3)
        Det 62  CVA-62   EA-1F (1)
Nov 64  Det 59  CVA-59   EA-1F (3)
Dec 64  Det 59  CVA-59   EA-1F (3)
Jan 65  Det 39  CVS-39   A-1E (1)
        Det 59  CVA-59   EA-1F (3)
Feb 65  Det 39  CVS-39   A-1E (1)
        Det 59  CVA-59   EA-1F (3)
Mar 65  Det 11  CVS-11   EA-1E (4)
Apr 65  Det 18  CVS-18   EA-1E (4)
        Det 39  CVS-39   EA-1E (4)
        Det 66  CVA-66   EA-1F (3)
May 65  Det 18  CVS-18   EA-1F (3), EA-1E (1)
        Det 39  CVS-39   EA-1E (4)
        Det 66  CVA-66   EA-1F (3)
Jun 65  Det 18  CVS-18   EA-1F (4)
        Det 21  Kindley  EA-1F (3)
        Det 39  CVS-39   EA-1E (4)
        Det 66  CVA-66   EA-1F (3)
Jul 65  Det 18  CVS-18   EA-1F (4)
        Det 66  CVA-66   EA-1F (3)
Aug 65  Det 66  CVA-66   EA-1F (3)
Sep 65  Det 66  CVA-66   EA-1F (1)
Oct 65  Det 16  Quonset  EA-1F (3), EA-1E (2)
        Det 66  CVA-66   EA-1F (5)
Nov 65  Det 16  Quonset  EA-1F (3), EA-1E (4)
        Det 66  CVA-66   EA-1F (4)
Dec 65  Det 18  CVS-18   EA-1E (2)
        Det 66  CVA-66   EA-1F (4)
Jan 66  Det 6   Rossie   EA-1F (3)
        Det 18  CVS-18   EA-1E (5)
        Det 42  CVA-42   EA-1F (3)
Feb 66  None
Mar 66  None
Apr 66  None
May 66  Det 3   Jax      EA-1F (4)
        Det 18  CVS-18   EA-1E (1)
May 66  Det 3   Jax      EA-1F (4)
        Det 18  CVS-18   EA-1E (1)
Jun 66  Det 3   Jax      EA-1F (4)
        Det 18  CVS-18   EA-1E (1)
Jul 66  Det 2A  Quonset  EA-1F (3)
        Det 3   Jax      EA-1F (4)
Aug 66  Det 2A  Quonset  EA-1F (3)
        Det 3   Jax      EA-1F (3)
Sep 66  Det 2A  Quonset  EA-1F (1)
        Det 3   Jax      EA-1F (3)
Oct 66  Det 2A  Quonset  EA-1F (1)
        Det 3   Jax      EA-1F (4), EC-1A (1)
        Det 66  CVA-66   EA-1F (1)
Nov 66  Det 2A  Quonset  EA-1F (1)
        Det 3   Jax      EA-1F (4), EC-1A (1)
        Det 66  CVA-66   EA-1F (3)
Dec 66  Det 2A  Quonset  EA-1F (1)
        Det 66  CVA-66   EA-1F (3)
Jan 67  Det 66  CVA-66   EA-1F (3)
Feb 67  Det 66  CVA-66   EA-1F (3)
Feb 67  Det 11  Quonset  EA-1F (3)
        Det 66  CVA-66   EA-1F (3)
Mar 67  Det 11  Quonset  EA-1F (3)
        Det 59  CVA-59   EA-1F (3)
        Det 66  CVA-66   EA-1F (3)
```

Mar 67	Det 11	Quonset	EA-1F (3)
	Det 59	CVA-59	EA-1F (3)
	Det 66	CVA-66	EA-1F (3)
Apr 67	Det 11	Quonset	EA-1F (3)
	Det 59	CVA-59	EA-1F (3)
	Det 66	CVA-66	EA-1F (3)
May 67	Det 11	Quonset	EA-1F (3)
	Det 59	CVA-59	EA-1F (3)
	Det 66	CVA-66	EA-1F (3)
Jun 67	Det 11	Quonset	EA-1F (3)
	Det 59	CVA-59	EA-1F (3)
	Det 66	CVA-66	EA-1F (3)
Jul 67	Det 11	Quonset	EA-1F (3)
	Det 66	CVA-66	EA-1F (3)
Aug 67	Det 11	Quonset	EA-1F (3)
Sep 67	Det 11	Quonset	EA-1F (3)
Oct 67	Det 11	Quonset	EA-1F (3)
Nov 67	Det 11	Quonset	EA-1F (3)
Dec 67	Det 11	Quonset	EA-1F (3)
	Det 14	CVA-14	EA-1F (3)
Jan 68	Det 14	CVA-14	EA-1F (3)

VAQ-33

Feb 68	Det 14	CVA-14	EA-1F (3)
Mar 68	Det 11	CVS-11	EA-1F (3)
	Det 14	CVA-14	EA-1F (3)
Apr 68	Det 11	CVS-11	EA-1F (3)
	Det 14	CVA-14	EA-1F (2)
May 68	Det 11	CVS-11	EA-1F (3)
	Det 14	CVA-14	EA-1F (2)
Jun 68	Det 11	CVS-11	EA-1F (3)
	Det 14	CVA-14	EA-1F (3)

Jul 68	Det 11	CVS-11	EA-1F (3)
	Det 62	CVA-62	EA-1F (3)
Aug 68	Det 11	CVS-11	EA-1F (3)
	Det 62	CVA-62	EA-1F (1)
Aug 68	Det 11	CVS-11	EA-1F (1)
	Det 62	CVA-62	EA-1F (1)
Sep 68	Det 11	CVS-11	EA-1F (3)
	Det 62	CVA-62	EA-1F (1)
Oct 68	Det 11	CVS-11	EA-1F (3)
	Det 62	CVA-62	EA-1F (1)
Nov 68	Det 11	CVS-11	EA-1F (3)
	Det 62	CVA-62	EA-1F (1)
Dec 68	Det 11	CVS-11	EA-1F (5)
	Det 62	CVA-62	EA-1F (1)
Jan 69	Det 11	CVS-11	EA-1F (5)
Feb 69	None		
Mar 69	Det 67	CVA-67	EA-1F (3)
Apr 69	Det 67	CVA-67	EA-1F (3)
May 69	Det 67	CVA-67	EA-1F (3)
Jun 69	Det 67	CVA-67	EA-1F (2)
Jul 69	Det 67	CVA-67	EA-1F (2)
Aug 69	Det 67	CVA-67	EA-1F (2)
Sep 69	Det 67	CVA-67	EA-1F (2)
Oct 69	Det 67	CVA-67	EA-1F (1)
Nov 69	Det 67	CVA-67	EA-1F (1)

Below, VC-33 AD-4Ns run-up for a deck launch off the USS Coral Sea (CVB-43) in 1952 (USN)

were lost or damaged in the Atlantic during this time. AD-3 BuNo 122840 stalled and crashlanded after making an attempted waveoff at the Millville Airport, NJ on 22 January 1951. The crew escaped without serious injuries. An AD-4N, BuNo 124144, ditched after takeoff from CVB-42 on 31 January 1951 with the crew safely exiting the sinking aircraft. The barrier on board CVL-49 saved AD-4N BuNo 124151 from total destruction during a night landing on 1 March 1951. Once more, the crew was uninjured. AD-3Q BuNo 122868 was forced to ditch after takeoff from CVB-43 on 2 March 1951. Again the crew escaped serious injury. A second VC-33 Skyraider from CVB-43, AD-4N BuNo 124132, flew into the water on 28 April 1951 after takeoff. Again the crew escaped safely. 15 September 1951 saw LCDR W.T. Morse Jr. go over the side in AD-2 BuNo 122225 during CarQuals after a failed emer-

gency landing due to a hydraulic failure. A pilot was killed on 30 November 1951 when he spun-in on takeoff from Atlantic City in AD-2 BuNo 122306. After the Leyte returned to the Atlantic on 14 January 1952, VC-33 AD-4Q BuNo 124042 bounced over the barrier on landing, cut off the tail of a VC-12 AD-3 and went over the side.

VC-33 received its first six AD-5Ns in April 1954. By 1 April 1955, all the AD-4N/NL were replaced by the AD-5Ns, but the AD-4Q/-3Qs lasted until 1 August 1956. On 2 July 1956, VC-33 was redesignated All-Weather Attack Squadron Thirty-Three VA(AW)-33 with the missions of night and all-weather attack, radar countermeasures, pathfinding and special weapons delivery. The Night Hawks became the largest carrier squadron in the world and had fifty AD-5Ns on-hand when the squadron became VA(AW)-33. A couple of shore-based S2F-2s and later TF-1Qs were operated as trainers from October 1956 to June 1969.

On 1 July 1957, the unit's tail code was changed from "SS" to "GD". In January 1958, six AD-5Qs were received and by 1 October a high of twenty-one AD-5Qs were assigned. The squadron's duty station was changed to NAS Quonset Point, RI, on 1 April 1958. On 30 June 1959, VA(AW)-33 was redesignated Carrier Airborne Early Warning Squadron Thirty-Three (VAW-33). In August

1961, twenty-two AD-5W Guppys were acquired from VAW-12 to help prosecute the squadron's new AEW/ASW mission.

In 1966, a VAW-33 Skyraider crewed by LTJG William Miller and right-seater LTJG Donald Erskine made the last landing on the Lake Champlain before it was decommissioned. The last AD-5W/EA-1E Det was aboard the Wasp in 1966 in support of the Gemini IX recovery. VAW-33 also participated in recovery of a

Above, three radar equipped VC-33 AD-5Ns in flight. Wing tip and tail trim was white. (USN) Below, VC-33 AD-5N BuNo 132561. Scrit under nose number was "Night Hawks". (Paul Minert collection)

Gemini vehicle in January 1965 from the Lake Champlain.

In 1967, in response to the

ond deployment was aboard the USS Intrepid (CVS-11) which was acting in a limited attack capability. The cruise was from 4 June 1968 to 8 February 1969. This deployment gave VAQ-33 the distinction of flying the Skyraider's last combat mission before retirement. During this cruise, AMS3 Gerald Feola was riding in the right seat of a Det 11 EA-1F in August when he was fired upon by a SAM. He related: "There was the usual small talk and chatter on the radio as the strike force assembled and aimed toward the coast, preparing for their strike mission. Our crew started jamming the enemy radar when, suddenly, we heard the foreboding transmission, "Launch, Launch, Launch!" All chatter stopped and matters grew serious. The pilot, LTJG Jenkins, quickly directed the crew to seek a "visual" on the SAM that was launched in our direction. Right before the call I snapped a picture of something I never saw before. It hap-

Above, VA(AW)-33 AD-5Q BuNo 134999, during the shakedown cruise of CVA-59. (Paul Minert collection) Bottom, VC-33 AD-5 BuNo 134998 taxis aboard the USS Intrepid (CVA-11) on 22 July 1956. Fin tip was red. (Naval History)

Vietnam War, VAW-33 was called

upon to supply two Dets to the fray. The first was from 27 December 1967 to 17 August 1968 aboard the USS Ticonderoga (CVA-14). During the deployment, on 1 February 1968, VAW-33 was redesignated Tactical Electronic Warfare Squadron Thirty-Three (VAQ-33). The squadron's nickname would also change from "Night Hawks" to "Firebirds". The sec-

VAW-33 NIGHT HAWKS

Above, VAW-33 EA-1E "Guppy" BuNo 139579 was assigned to the USS Wasp (CVS-18), seen here at an open house demonstrating its ability to act as an attack aircraft by mounting a bomb on the inner wing pylon. (William Swisher) At right, VAW-33 AD-5W "Guppy" BuNo 132778 aboard the USS Essex (CVS-9). Wing tips and rudder trim were maroon. (Ginter collection) Bottom, VAW-33 Det 43 AD-5Q BuNo 132575 from the USS Saratoga (CVA-60) in December 1961. (USN)

VAW-33

Above, VAW-33 EA-1F BuNo 132543 leaves the deck of USS Forrestal (CVA-59). Wing tips and rudder trim were maroon. (Paul Minert collection) At left, VAW-33 EA-1F BuNo 132618 in flight from the USS Independence (CVA-62) on 1 May 1962. Tail trim was maroon. (USN) Below, VA(AW)-33 AD-5N BuNo 132609 taxiing with its new 1 July 1957 "GD" tail code. (Paul Minert collection)

Above, VAW-33 EA-1F BuNo 132521 near NAS Quonset Pt. RI, on 6 September 1962. (USN) Bottom, VAW-33 EA-1Fs being loaded aboard the USS Independence (CVA-62) on 6 August 1963. Wing tips and rudder trim were maroon. (USN)

VA(AW)-33/VAQ-33

Above, VA(AW)-33 A-1E BuNo 132614 runs-up at NAS Alameda, CA, in July 1965. Aircraft had previously been assigned to CVW-7 and is seen here being returned to service, presumably to be transferred to the USAF or the VNAF. (William Swisher) At left, VAW-33 A-1E BuNo 133907 target tug with tow gear installed on the centerline. Aircraft was finished in utility colors of engine grey with yellow wings and horizontal tail surfaces and da-glo red tail and wing stripe. (Paul Minert collection) Bottom, VAW-33 EA-1F BuNo 132572 from the USS Ticonderoga (CVA-14) taxis at NAS Atsugi, Japan, on 19 January 1968. (Toyokazu Matsuzaki)

Above, Ticonderoga-based VAQ-33 EA-1F BuNo 134974. (Paul Minert collection) Below, VAQ-33 EA-1E BuNo 135188 in February 1967 had a black and white checked engine cowl. The drop tanks were red on top, dark blue on bottom with a yellow lightning bolt down the middle. A small unit logo is found on the drop tank noses. Tail stripe and wing tip stripes were black with yellow lightning bolts. (Paul Minert collection)

pened to be the exhaust trail of the SAM that was rising from its launch pad. Shortly, Jenkins banked the EA-1F hard right toward the SAM! I asked myself, "What's he doing!" Turns out it was an effective evasive maneuver. I remember vividly the SAM speeding by us. It resembled a telephone pole with small fins. To

this day, I still see the image of the impulse trail from its rocket engine, leaving what looked like dashes in the sky behind it. That baby was booking! Thankfully, the missile missed its mark because it was being jammed (confused) by those wonderful electronic countermeasure (ECM) guys in the back seats. The SAM plunged into the ocean.

The strike group proceeded to its target and bombed it accordingly, then reversed course toward the sea."

VAQ-33's last Skyraider Det was Det 67 aboard the newly commissioned USS John F. Kennedy (CVA-67) from 5 April to 21 December 1969. The OIC for the last harah was LCDR D.R. Quinn. During the cruise, one EA-1F was lost to an engine failure on takeoff and the crew including one midshipman were safely rescued. By February 1970, the last Skyraider was retired.

"Ode from a SPAD"

I am old, I am weary, I am worn and I'm tired. Yet My job I will do with the greatest of pride. History will tell of my famous deeds, and the men who have flown me, I wish them God's speed. For the past twenty years, I have served in the sky, and now it is my turn to fade out and die. I'm not so fast, I'm not so sleek, just a small part of fat. I'm a "SPAD" and that makes me the "Greatest of Great".

Above, VAQ-33 EA-1F from CVS-11, like the one attacked by a SAM, prepares to launch on a jamming mission over Vietnam. (USN) Below, VAQ-33 EA-1F BuNo 132506 AB/752 made the last Skyraider takeoff from the USS John F. Kennedy (CVA-67) on 20 December 1969. (Ginter collection) Bottom, VAQ-33 EA-1F BuNo 132575 at NAS Quonset Pt, RI, on 22 June 1968. (via Tailhook)

ATTACK SQUADRON THIRTY - FOUR, VA-34 "BLACK PANTHERS"
SECOND ATTACK SQUADRON THIRTY - FIVE, VA-35 "BLACK PANTHERS"

The second Attack Squadron Thirty-Five (VA-35) was first established as Bombing Squadron Three B (VA-3B) on 1 July 1934 at NAS Norfolk, VA. Equipped with Martin BM-1/2s, VA-3B was redesignated Bombing Squadron Four (VB-4) on 1 July 1937. In January 1938, VB-4 received the Vought SB2U Vindicator prior to being redesignated VB-3 on 1 July 1939. VB-3 began operating the BT-1 and SBC-4 in March 1941, the SBD-3 in August 1941, the SBD-4 and SBD-3P in April 1943, the SBD-5 in August 1943, the SB2C-1C in December 1943, the SBW-3 in July 1944, the SB2C-4 in September 1944, and the SB2C-5 in January 1946. Then, on 15 November 1946, VB-3 was redesignated Attack Squadron Three A (VA-3A) before becoming Attack Squadron Thirty-

Four on 7 August 1948.

While under the command of LCDR Heber J. Badger, VA-34 received its first AD-2 on 24 November 1948 while aboard the USS Kearsarge (CV-33) as part of CVG-3. Then, on 30 November 1948, LCDR Ralph M. Bagwell relieved LCDR Badger and by the end of December 1948, VA-34 had fifteen AD-2s and ten SB2C-5s at Charlestown. In February 1949, VA-34 boarded the USS Coral Sea (CVB-43) with sixteen AD-2s, one AD-2Q and one SB2C-5 for carrier qualifications. The AD-3 started replacing the AD-2s in November 1949 with twelve AD-3s and 16 AD-2s on-hand at NAS Quonset Pt., RI, at the end of the month. VA-34 had seventeen AD-3 on strength on 15 February 1950 when VA-34 was redesignated VA-35.

The squadron's first Skyraider

deployment was to the Mediterranean aboard the USS Leyte (CV-32) from 2 May through 24 August 1950. On 6 September, CV-32 was transferred to the Pacific due to the Korean War via the Panama Canal, arriving in San Diego on 18 September. CV-32 was ordered to Korea on 19 September with eighteen AD-3s, which were supplemented with three AD-4s in October. The squadron's first combat mission was launched on 11 October and on 12 December 1950 LCDR Bagwell crash landed in North Korea and was taken prisoner. Seven days

---

Above, VA-34 AD-2Q in flight on 28 November 1949. Note squadron insignia on the cowl. (via Tailhook) Below, two VA-34 AD-2s aboard the USS Leyte (CV-32), bombed-up with three torpedoes and twelve rockets for a fire power demonstration on 4 May 1949. (National Archives via Jim Sullivan)

Above, VA-35 AD-3 BuNo 122799 taxis on the USS Leyte (CV-32) on 11 November 1950. Pilot was ENS R. Kissimey. (USN via Tailhook) At left, VA-35 AD-3 K/513 folds its wings after trapping aboard CV-32 on 20 November 1950. Deck crew hustles a live rocket that was dropped on landing to the deck edge before tossing it in the sea. Fin tip was green. (National Archives via Jim Sullivan) Below, VA-35 AD-3 piloted by ENS William C. Pierson lands aboard CV-32 on 13 November 1950. (USN via Tailhook)

Three views of VA-35 AD-2 K/505 piloted by ENS D.F. Kirkpatrick. It clipped its wing on the 5" gun turret on landing aboard the USS Leyte (CV-32) on 10 November 1950, then proceeded over the side clawing for air before settling on the flat, calm sea. In the last photo the pilot can be seen walking across the wing. (National Archives via Jim Sullivan)

later, LCDR John G. Osborn took command of VA-35. LTJG Roland R. Batson, Jr. also crashed landed and became listed as missing-in-action.

For its actions while deployed to Korea, VA-35 was awarded the Navy Unit Commendation and the Korean Presidential Unit Commendation. The squadron returned to Quonset Pt. on 3 February 1951 and was equipped with AD-4/4Ls for a short deployment to the Med from 3 September to 21 December 1951 aboard Leyte. On the squadron's return to CONUS, it was assigned to NAAS Sanford, FL. From there, VA-35 made one more deployment aboard Leyte, again to the Med,

from 29 August 1952 through February 1953.

VA-35 was transferred to NAS Cecil Field, FL, in February 1953 where they received AD-4Bs and AD-4Ns before acquiring AD-6s in September 1953 (the first fleet squadron to equip with AD-6s). From here, VA-35 deployed to the Med aboard the USS Tarawa (CVA-40) from 12 November 1953 through September 1954 for a World Cruise. In November 1954, CDR C. W. Johnson relieved LCDR David G. Adams, Jr. as CO of VA-35.

CVA-14, the USS Ticonderoga,

hosted VA-35 on its next Med Deployment from 4 November 1955 through 2 August 1956. On return to Cecil Field, CDR Alfred E. Brown took over as the squadron's CO.

For the next six deployments, VA-35 was assigned to the new super carrier the USS Saratoga (CVA-60). All deployments were to the Med with the third also visiting the North Atlantic. During the first CVA-60 deployment from 1 February to 1 October 1958, VA-35 supported the Marine landings in Lebanon. On return to CONUS, the squadron was transferred to NAS Jacksonville, FL, where CDR Harlan W. Foote relieved CDR Brown on 7 October 1958. The second Saratoga cruise was from 16 August 1959 to 26 February 1960; the third with CDR William F. Bailey at the helm was from 22 August 1960 to 26 February 1961. On 28 April 1961, CDR E. C. Hastings II took command and the squadron's fourth Saratoga

At top left, VA-35 BuNo 139799 runs-up aboard Saratoga. Prop hub and tail stripes were orange. (USN) Below, VA-35 A-1Hs in flight: AC/404 BuNo 134501, AC/403 BuNo 134535, AC/401 BuNo 139799 and AC/402 BuNo 137564. (USN)

deployment was from 28 November 1961 to 11 May 1962. CDR W. F. Walker became CO on 29 May 1962 but was relieved by CDR J. R. Constantine on 7 March 1963, who skippered VA-35 for its fifth deployment aboard CVA-60 from 29 March through 25 October 1963. For the sixth Saratoga cruise from 28 November 1964 until 12 July 1965, CDR J. B. Allred was in command until 22 January when CDR Richard G. Layser took over the helm. Layser died in an accident two weeks later, on 4 February, and in turn was temporarily replaced by LCDR Joseph F. Frick. CDR John W. Shute took over on 22 March in time to finish the deployment.

During the deployment, two pilots, LCDR Jerry Loeb and LTJG Ronald "Banty" Marron, were interned in Algeria. On 4 January 1965, they launched on a search mission to find the Italian fishing vessel, the Leopold II, that had sent out an SOS. The estimated distance to the search area was 260nm in deteriorating weather. An A-3, supposed to act as a communication relay to the Saratoga and the USS Little Rock (CL-92), was detached to proceed to the area. While enroute, the Leopold had cancelled the SOS and was under tow to safety. This fact and the cancellation of the A-3 and the Little Rock were never relayed to the A-1 pilots, and CVA-60 had gone radio silent. As the two pilots searched the area, the weather worsened to the point that Banty's prop governor had frozen, so to avoid a ditching at sea they climbed and headed for land in Muslim Algeria. They both made a successful landing on a snow covered plateau near Djelfa, and were immediately arrested by the military and incarcerated in Djelfa. A few days later they were moved to Algiers, but their survival gear, knives, smoke flares, flare guns and cartridges were not taken from them. So a few nights later, they made an escape into the rainy night. They then took a cab driver hostage at knifepoint and had him deliver them to the US embassy. He was subsequently killed by the

Above, VA-35 AD-6 BuNo 139799 on the deck of the USS Saratoga (CVA-60). (USN) Below, VA-35 AD-6 BuNo 135381 assigned to the USS Saratoga (CVA-60) in 1959-60 with Mk. 76 practice bombs. (William T. Larkins)

Algerian Army, who slit his throat and blamed it on the Americans. What followed was a dangerous and convoluted political dance that almost landed the pilots back in prison. Eventually, they boarded an Air France flight and made it to Marseille and finally the ship. Their return was not a celebration, but a painful debrief, with a JAG investigation and Pilot Disposition Board followed by a gag order.

On return from the squadron's last Skyraider cruise, the squadron was transferred to NAS Oceana, VA. The squadron had left two aircraft in Algeria, and lost two more shortly after Loeb's and Marron's return leaving them with only five Skyraiders when they arrived stateside.

On 15 September 1965, the squadron started transitioning to the A-6A Intruder, which began combat operations in Victnam in November 1966. The last Skyraider was trans-

Above, VA-35 A-1H BuNo 135258 on Saratoga in March 1964. Tail stripes were orange. (USN) Below, VA-35 AD-5 BuNo 132444 on 8 March 1962 in utility squadron colors of engine grey fuselage, yellow wings, and international orange vertical tail and wing stripes. The aircraft is fitted with the tow target kit on its belly. (USN)

ferred out in October 1965.

COMPOSITE SQUADRON THIRTY - FIVE, VC-35
ATTACK SQUADRON (ALL - WEATHER) THIRTY - FIVE, VA(AW)-35
ATTACK SQUADRON ONE - TWENTY - TWO, VA-122

Composite Squadron Thirty-Five (VC-35) was established at NAS San Diego, CA, on 25 May 1950 with attack assets from VC-3, the Pacific Coast nocturnal fighter/attack squadron. VC-3 would continue to provide night fighter detachments to the fleet and VC-35 would provide night attack detachments.

One month later, we entered the Korean War with only three Essex class carriers in the Pacific. VC-35 only had two AD-4Ns and one AD-3Q on-hand when the war broke out while VC-3 still had eight Skyraiders with F4U/AD detachments being sent aboard CV-45 and CV-47. By the end of July, there were eight Skyraiders on-hand and by September there were forty-one Skyraiders assigned. These were: fifteen AD-4Ns, eight AD-4Qs, two AD-3s, six AD-3Ns, three AD-2Qs, and seven AD-1Qs.

The carrier detachments VC-35 provided were also called night attack or VAN teams. At first, these teams or detachments were assigned numbers, the first being Team 3, but by the end of 1951 they were designated by a letter, the first being Det or Team C. After the change, the squadron continued to track their Dets as numbers, too, the last being Team 63 (Det L) in 1959.

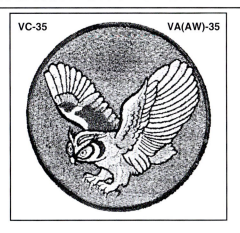

VC-35   VA(AW)-35

VC-35's first Det to deploy was aboard the USS Boxer (CV-21) from 24 August to 11 November 1950 as part of CVG-2 with four AD-4Ns. CVG-2 had hastily transferred from the East Coast in time to replace CVG-19 for the cruise. This included a VC-33 Det, whose planes and personnel were absorbed into VC-35.

The second Det was Team 3, commanded by LT Franklin Metzner aboard the USS Princeton (CV-37) from 9 November 1950 to 11 June 1951. Combat operations commenced on 12 December 1950, with VC-35 flying antisubmarine patrol (ASP) and courier flights through 12 December. Night heckler missions began on the 13th with LT Metzner and ENS J.D. Ness destroying five trucks and attacking the Chinese-held village of Sahong-Ni. The heckler

VA-122

missions continued through 24 December in support of the evacuation of Hamhung, during which seventeen night missions were flown and sixteen trucks and an ammunition dump were destroyed. Then, three ASP missions finished out the year on the 28th. After a short R&R in Sasebo, CV-37 returned to the line on 19 January 1951 and Det 3 returned to night heckling. On the 24th, LT Metzner's AD was hit by ground fire and LT English was hit the next day. On 1 February, LT Atlee Clapp and LTJG Rodgers caught a locomotive in the open and destroyed it. This was

---

Below, VC-35 AD-1Q BuNo 09359. All the white lettering has been over-sprayed with a black tint. (William Swisher)

the first in a long series of train tunnel busting missions and was flown on 9 February.

Three VC-35 pilots loaded their AD-4Ns with torpedoes and took part in the Hwachon Dam attack on 30 April 1951. The three pilots, LT Atlee Clapp, OIC LT Frank Metzner, and LT Addison English, were in division one with CAG-19 CDR Richard Merrick in his AD-4Q. VA-

Above, two VC-35 AD-2Qs near San Diego with BuNo 122387 in the foreground in August 1950. (National Archives) Below, flight of four VC-35 Skyraiders over San Diego on 4 October 1950 with AD-3 BuNo 122748 in front followed by three AD-3Qs. (National Archives)

VC-35 / VA(AW)-35 / VAW-35 DETACHMENTS (DETS)

DET	SHIP	AIR GROUP	ADMIN SQD	DEPLOYED DATES	O-in-C
?	CV-21	2	VA-65	24 Aug. 1950 to 11 Nov. 1950	LT W.T. Snipes
3	CV-37	19	VA-195	09 Nov. 1950 to 11 Jun. 1951	LT F. Metzner
4	CV-45	2	VA-65	06 Dec. 1950 to 29 Mar. 1951	LT M.E. Beaulieu
4	CV-47	2	VA-65	29 Mar. 1951 to 09 Jun. 1951	LT M.E. Beaulieu
5-A	CV-21	101	VA-702	02 Mar. 1951 to 25 Oct. 1951	LT D. Arrivee KIA
					LTJG W.C. Raposa
6-G	CV-31	102	VA-923	10 May 1951 to 17 Dec. 1951	LCDR A.C. Wallman
7	CV-37	19X	VA-55	29 May 1951 to 29 Aug. 1951	LT A. Borysiewicz
8-B	CV-9	5	VF-54	26 Jun. 1951 to 25 Mar. 1952	LCDR F. Bertagna
9-D	CV-36	15	VA-728	08 Sep. 1951 to 02 May 1952	LT R.C. Bartlett
10-H	CV-45	ATG-1	VF-194	15 Oct. 1951 to 03 July 1952	LT M. Schluter
C	CV-47	11	VA-115	31 Dec. 1951 to 08 Aug. 1952	LT F.D. Hooks
A	CV-21	2	VA-65	08 Feb. 1951 to 26 Sep. 1952	LT R.W. Taylor
E	CV-37	19	VA-195	21 Mar. 1952 to 03 Nov. 1952	LT R.L. Bothwell
I	CV-9	ATG-2	VA-55	16 Jun. 1952 to 06 Feb. 1952	LCDR E.H. Potter
W	Atsugi			16 Jun. 1952 to 15 Sep. 1953	CDR W. Conley
F	CV-33	101	VA-702	11 Aug. 1952 to 17 Mar. 1953	LCDR M. Brambilla
G	CV-34	102	VA-923	15 Sep. 1952 to 18 May 1953	LT W.P. Kiser
B	CVA-45	5	VF-54	20 Nov. 1952 to 25 Jun. 1953	LCDR W. C. Griese
M	CVA-47	9	VA-95	15 Dec. 1952 to 14 Aug. 1953	LCDR F. Ward
D	CVA-37	15	VA-155	24 Jan. 1953 to 21 Sep. 1953	LT J.C. Holloway
H	CVA-21	ATG-1	VF-194	30 Mar. 1953 to 28 Nov. 1953	LT C.R. Johnson
C	CVA-33	11	VA-115	01 July 1953 to 18 Jan. 1954	LCDR McCartney
A	CVA-10	2	VA-65	03 Aug. 1953 to 03 Mar. 1954	LCDR J. Gerry
E	CVA-34	19	VA-195	14 Sep. 1953 to 22 Apr. 1954	LCDR D.K. Gibbs
I	CVA-9	ATG-2	VA-55	01 Dec. 1953 to 12 July 1954	LCDR A.N. Nelson
F	CVA-15	14	VA-145	03 Feb. 1954 to 06 Aug. 1954	LCDR J. Knosp
G	CVA-21	12	VA-125	03 Mar. 1954 to 11 Oct. 1954	LCDR W.N. Nelson
B	CVA-47	5	VF-54	12 Mar. 1954 to 18 Nov. 1954	LCDR C.A. Hill Jr.
M	CVA-12	9	VA-95	11 May 1954 to 10 Dec. 1954	LCDR M.S. Essary
D	CVA-10	15	VA-155	01 July 1954 to 28 Feb. 1955	LT T.M. Taylor
H	CVA-18	ATG-1	VF-194	01 Sep. 1954 to 11 Apr. 1955	LCDR E.W. Gendron
C	CVA-33	11	VA-115	07 Oct. 1954 to 12 May 1955	LCDR H. O'Connor
A	CVA-9	2	VA-65	03 Nov. 1954 to 21 Jun. 1955	LCDR J. Gallagher
E	CVA-34	19	VA-195	02 Mar. 1955 to 21 Sep. 1955	LCDR W.F. Bailey
I	CVA-47	ATG-2	VA-55	01 Apr. 1955 to 23 Nov. 1955	LT D. Munns
F	CVA-21	14	VA-145	03 Jun. 1955 to 03 Feb. 1956	LT R.F. Smith
G	CVA-19	12	VA-125	10 Aug. 1955 to 15 Mar. 1956	LT W.B. Muncie
B	CVA-33	5	VF-54	29 Oct. 1955 to 17 May 1956	LT Greig
J	CVA-38	ATG-3	VF-92	05 Jan. 1956 to 23 Jun. 1956	LT W.P. Robinson
M	CVA-34	9	VA-95	18 Feb. 1956 to 13 Aug. 1956	LCDR E. Mullkoff
K	CVA-10	ATG-4	VA-216	19 Mar. 1956 to 13 Sep. 1956	LT R.F. Stelloh
D	CVA-18	15	VA-155	23 Apr. 1956 to 15 Oct. 1956	LT R.E. Forbis
H	CVA-16	ATG-1	VA-196	28 May 1956 to 20 Dec. 1956	LCDR J.E. McBride
C	CVA-9	11	VA-115	16 July 1956 to 26 Jan. 1957	LCDR B.F. Jones
L	CVA-31	21	VA-215	16 Aug. 1956 to 28 Feb. 1957	LCDR E. McGonigle
A	CVA-38	2	VA-65	13 Nov. 1956 to 20 May 1957	LCDR G. Richardson
F	CVA-12	14	VA-145	21 Jan. 1957 to 25 July 1957	LCDR M. Polkowske
E	CVA-10	19	VA-195	09 Mar. 1957 to 25 Aug. 1957	LCDR J. Hennessy
I	CVA-19	ATG-2	VA-55	06 Apr. 1957 to 18 Sep. 1957	LT H. Howser
G	CVA-16	12	VA-125	19 Apr. 1957 to 17 Oct. 1957	LT D.A. Woodward
B	CVA-31	5	VF-54	12 July 1957 to 09 Dec. 1957	LT E. McCullum Jr.
J	CVA-33	ATG-3	VA-96	09 Aug. 1957 to 02 Apr. 1958	LCDR A. McNeill
M	CVA-14	9	VA-95	16 Sep. 1957 to 25 Apr. 1958	LCDR B. Jones
K	CVA-12	ATG-4	VA-216	08 Jan. 1958 to 30 Jun. 1958	LT J. Stackpole
D	CVA-19	15	VA-155	16 Feb. 1958 to 03 Oct. 1958	LT J.T. French
C	CVA-38	11	VA-115	00 Mar. 1958 to 22 Nov. 1958	LCDR H. Howser
L	CVA-16	21	VA-215	17 July 1958 to 17 Dec. 1958	LT F.P. Davis
A	CVA-41	2	VA-65	16 Aug. 1958 to 12 Mar. 1959	LT F.L. Nelson
K	CVA-20	ATG-4	VA-216	21 Aug. 1958 to 12 Jan. 1959	LCDR M. Polkowske
H	CVA-14	ATG-1	VA-196	04 Oct. 1958 to 16 Feb. 1959	LCDR W.H. Cleland
E	CVA-31	19	VA-195	01 Nov. 1958 to 18 Jun. 1959	LCDR G. Johnson
F	CVA-61	14	VA-145	03 Jan. 1959 to 27 Jul. 1959	LCDR C. Gallagher
C	CVA-38	11	VA-115	09 Mar. 1959 to 03 Oct. 1959	LT G.C. Lyne
L	CVA-16	21	VA-215	26 Apr. 1959 to 01 Dec. 1959	LCDR L. Snider

Det E — VAN 13

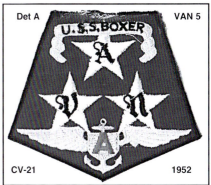

Det A — VAN 5 — CV-21 — 1952

Det I — VAN 14

Det M — CV-47 — 1952

Above, VC-35 AD-4NL BuNo 124745 in flight with underwing searchlight on 6 October 1951. (National Archives) Below, VC-35 AD-4NL BuNo 124741 with APS-31B radar pod in flight on 6 October 1951. Note extended 20mm gun barrels with flash suppressors for night operations. (National Archives)

195's LCDR Carlson, ENS Robert Bennett, LTJG E.V. Phillips, and LTJG J.R. Sanderson made up the second division. The strike totally destroyed one floodgate and blew away the bottom portion of another, expelling water into the river below and thereby denying the enemy control of the dam's water system (see pages 87-88).

On 11 May 1951, OIC LT Metzner and CAE Richard Green completed a bridge strike near Pyongyang and instead of returning to the ship landed at K-18 and reloaded with bombs and ammo. On take-off, after reaching about 250 ft, lift was lost and the AD-4N crashed, shedding its wings and bombs, cartwheeled and burned. Both crewmen escaped with

minor injuries, but the cause of the accident was what was interesting. They had taken off with the wings folded! Another Det 3 aircraft, AD-4Q BuNo 124058, was lost on 9 June 1951 due to AAA fire.

The third VC-35 Det, Team 4 aboard the USS Valley Forge (CV-45), arrived on station in late December 1950 and flew their first combat mission on the night of the 23rd. The night heckler mission was flown by OIC LT B.E. Beaulieu and ENS J.F. Bachman. Team 4 transferred to the USS Philippine Sea (CV-47) on 29 March 1951 and was deployed through 9 June 1951. Det 4's OIC on both ships was LT Beaulieu.

The fourth Korean War Det was Team 5A aboard the USS Boxer (CV-21) from 2 March to 25 October 1951. The OIC was LT D. Arrivee who was killed-in-action and replaced with LTJG W.C. Raposa. Arrivee was killed on 21 June while flying an armed reconnaissance mission near Sagi-ri in a VA-702 Skyraider.

VC-35's next Korean War Det was Team 6 or Det G aboard the USS Bon Homme Richard (CV-31), with OIC LCDR A.C. Wallman from 10 May to 17 December 1951. LT W.H. Roundtree ditched twice after takeoff in AD-4Ns on 7 July and on 11 July (BuNo 124137) 1951. On 3 October, LTJG Robert Probyn's aircraft was hit by ground fire, but made it to the water before the engine quit and he ditched. On 4 November, ENS Gerry Canaan was hit by flak, bailed out and was captured.

Team 6G was followed by Team 7 off Korea aboard the USS Princeton (CV-37) from 20 May to 29 August 1951. Combat operations began on 2 June and on 10 June CAG 19X, CDR Charlie Stapler (former CO of VC-35), was lost on a heckler mission in a Det 7 Skyraider. Stapler went down with the aircraft but AT-1 R.L. Blazevic bailed out and was captured. Team 7's OIC LT Borysiewicz and LTJG Oliver destroyed six trucks on the night of 30 June. Det 7 scored four trucks destroyed and eight damaged

on 18 July, seventeen trucks were put out of commission on 20 July, and ten more trucks were destroyed on the night of 5 August.

Team 8B aboard the USS Essex (CV-9) was the next Det to join the war effort off Korea. They operated from 26 June 1951 to 25 March 1952. LTJG L.D. Smith and ATAN Phillip Batch were killed when their bombed-up aircraft exploded after launch on 26 August. Another aircraft flown by LTJG Hessom was hit by a 37mm round which put a man-sized hole in his wing. Despite the damage, he made it back to the ship for a safe landing.

Team 9 or Det D was the next VC-35 combat Det. They operated off the USS Antietam (CV-36) from 8 September 1951 to 2 May 1952. On 4 November, LTJG N.K. "Moose" Donahoe, AM3 R.A. Nobles and AT3 J.A. Beecher ditched after a bad cat shot in their AD-4NL. All were rescued by the plane guard, USS Uhlmann (DD-687). Donahoe lost another AD-4NL on 13 December. He was hit in the wing by a 37mm shell during a night heckler mission and subsequently crash-landed at K-18 where the aircraft was written off.

ATG-1 with Team 10 or Det H was aboard CV-45 from 15 October 1951 to 3 July 1952. Combat operations began on 11 December and on the 13th LTJG Harry Ettinger was shot down by a 37mm round on his first combat mission. He and AT2 J.H. Gilliland and AOU3 J.R. McElroy bailed out and were captured by the North Koreans. Another aircraft, an AD-4NL, was hit by AAA on 27 April 1952 but was able to ditch in Wonsan Harbor. LT Wayne Shepard and AD3 D.F. Lovell were recovered safely.

Eleven more deployments were made by VC-35 Dets to Korea before the war ended on 27 July 1953. More aircraft losses and crew injuries followed: AD-4N BuNo 124731 stalled on landing aboard CV-47 off NAS San Diego on 5 December, 1951, during work-ups for the carriers second war cruise. The pilot survived the crash without major injuries. Det E's

VC-35 DETS

DET B   1952-53

DET C   CV-47

1952

DET G   1952-53

Above, bombed-up VC-35 Det C AD-4NL BuNo 124748 low over the water inbound to Korea from the USS Philippine Sea (CV-47) in April 1952. Note squadron insignia below the windscreen. (National Archives) At left, bombed-up VC-35 Det. H AD-4N NR/89 taxis off the elevator on the USS Boxer (CVA-21) in June 1953. (National Archives) Below, VC-35 Det C AD-4NL launches off the USS Philippine Sea (CV-47) in March 1952. (National Archives)

LT R.E. Garver and AT-3 A.F. Ruddell were killed on 8 June 1952

when they crashed into the sea after an armed weather flight. Det A lost two crewmen, AT3 W.B. Burdette and AT-2 B.G. Soden, to a hangar deck fire on 6 August 1952 aboard Boxer. AD-4N BuNo 125710 was hit by AA and ditched near Yang-do on 8 August 1952. LT J.C. Norton, AT2 B.B. Killingsworth and AN J.A. Stephens were safely recovered. A Det F AD-4N piloted by LT Francis Anderson was hit by a rifle bullet which grazed AT3 Schmid near his left eye on 18 September 1952. After landing, he was evacuated to Yokosuka where he recovered. He returned to Det F and was flying again with Anderson when on 28 January 1953 they failed to return from a night heckler mission. AD-4N BuNo 125712 ditched after takeoff from CVA-33 on 5 October 1952. AD-4N BuNo 125709 vanished during a night mission on 28 January 1953. AD-4N BuNo 126944 ditched off CVA-21 on 5 July 1953. LT C.R. Johnson and AOAN D.G. Kennedy were rescued by DD-692 but AO1 M.J. Wright was lost. VC-35's final losses of the war were on 14 July 1953 when LT R.A. Smith, AFAN J.S. Kennedy and ADAN T.H. Guyn failed to return to Boxer.

A dedicated night heckler operation against railroads called "Moonlight Sonata" began on 15 January 1952. During the period, up to five two-plane sections would cover a 50-mile section of track each on flyable nights. Two locomotives were destroyed and three were damaged. In May Operation "Insomnia" was conducted which destroyed nine trains and heavily damaged two.

Destruction of truck convoys continued in spite of the concentration on the railroad. On 18 April 1952, LT A.R. Kreutz and ENS P.J. Weiland from CV-21 destroyed eight trucks, two warehouses and an ammo dump as well as damaging twelve other trucks. On the 23rd, two other CV-21 pilots, LT C.H. Hutchinson and LT D.G. Creeden, destroyed a bridge and four trucks, burned a warehouse, and damaged six trucks. From 15 July to 3 August 1952, twenty-three trucks were destroyed and sixty-eight

were damaged as well as two trains being destroyed. Another train was destroyed on 4 October, another on the 5th and one more by the end of the month.

A number of special weapons AD-4Bs were assigned to VC-35 and a special Det, Det W, was formed and maintained at NAF Atsugi from 16 June 1952 to 15 September 1953.

From 28 August to 2 September 1952, Det A aboard CV-21 had two AD-2Qs configured as drone control aircraft used for the combat evaluation of the F6F-5K assault drone fitted with a 2,000lb bomb. The Hellcat's first strike was against the Hamhung railway bridge which resulted in a near miss due to controller error. The second mission against the Chungjin railway bridge resulted in a direct hit on a bridge pylon. The 3rd and 4th mission on a railroad tunnel and a hydroelectric plant were near misses with little or no damage. The 5th strike failed due to equipment failures on the F6F, and the results of the last strike against a railway bridge were not known because of a TV system failure. The program was terminated after the test and no further use of drones in combat followed.

1953 was offensively much like 1952 as far as night heckler and interdiction missions were concerned. A new ground attack weapon was added to VC-35's inventory with the arrival of CVA-37 for its third war cruise in February 1953. Det M was tasked with the operational evaluation of the podded 2.75-inch "Mighty Mouse" folding fin rockets. Originally an air-to-air weapon, it was evaluated by VC-35's XO, CDR Frank G. Edwards, at NOTS Inyokern (see NF98 pages 182-186) in December 1952 for ground targets. On the night of 13 April, Det M's OIC LCDR Felix Ward and crewmen AD3 E.B. Willis and AD2 R.M. Yonke encountered a truck convoy west of Wonsan. When he rolled in on his first pass, the lead truck stopped at the bridge. Ward didn't fire but made several more passes during which all the trucks (at least 20) bunched up on the leader. He then dropped a flare and made a fir-

DET H   VAN 21

CVA-21   1953

DET A   VAN 23

CVA-10   1954-55

DET E   VAN 24
CVA-34   1953-54

DET I   VAN 25

CVA-9   1953-54

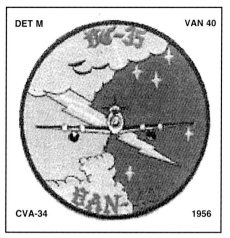

ing pass. He unloaded all six of his rocket pods on the column destroying most of the trucks and setting off secondary explosions. This success was repeated on the 26th with LT P.E. Sullivan destroying five plus trucks with two pods; LT E.R. DeSmet expended three pods and disposed of four trucks and damaged two more, then with his remaining three pods he destroyed an AA site and three to four buildings; LT L.W. Erickson burned two buildings and destroyed an AAA site with his rockets; and OIC Ward burned half the village of Soho-ri in a four pod attack.

After the 27 July 1953 armistice, VC-35 operations continued at a hectic pace. Unlike the systematic reduction of carriers after WWII, carrier strength continued to grow after Korea. The end of the war did allow for a more formal training cycle for the Dets, which did not short-suit any needed skills. The training for a Det began five months prior to deployment and was conducted at San Diego, except for five weeks spent at NAAS El Centro, CA, on day and night weapons training. Training included ASW, ECM, mine-laying, special weapons delivery, night heckler missions, and day and night carrier qualifications.

Above, two VC-35 AD-2Q drone control aircraft take-off on a mission from CVA-21 after the launch of their F6F-5K assault drone on a strike over Korea. (USN) Below, VC-35 AD-5N conducting FCLP at NAAS Brown Field on 28 May 1954. (USN)

In March 1954, VC-35 received their first six AD-5Ns. By May they had two AD-5s, fourteen AD-5Ns, nine AD-4Ls, twenty-six AD-4Ns, three AD-4Qs, one AD-3, one AD-3Q, and two AD-2Qs as well as the squadron's first two AD-6s. The AD-5N became the dominant version with fifty-two on-hand by the end of December 1954.

In November 1955, an F3D-2 SkyKnight was received for evaluation and in March 1956 a second SkyKnight was acquired. The F3Ds were operated through February 1957. In October 1956, an S2F-1 was received, then a TF-1Q in November and another in March 1957. They were operated through May 1959. The first AD-5Q was added in November 1957. A high of seventeen were on-hand in October 1958 and the type was withdrawn in March

1959.

VC-35 was redesignated All-Weather Attack Squadron Thirty-Five (VA(AW)-35) on 1 July 1956. A new mission was also added, the training of fleet replacement pilots and air-crews for the Skyraider. With this mission came the slow elimination of the squadron's other missions, with the last fleet Det, Det L, returning on 1 December 1959. With the return of Det Lima, VA(AW)-35 was redesignated Attack Squadron One-Two-Two (VA-122) on 30 June 1959 and

assigned to Replacement Air Group Twelve (RCVG-12). A large number of AD-6/7s were added quickly as were T-28Bs. By 1 January 1960, there were seven AD-7s, seventeen AD-6s, two AD-5s and ten T-28Bs on-hand.

The AD RAG was transferred to NAS Moffett Field, CA, in July 1961 and finally to NAS Lemoore, CA, in February 1963. In November 1966, eight South Vietnamese Air Force officer pilots were trained on the AD-6/7. Novernber 1966 also marked the

Above, VC-35 AD-4N BuNo 125740 at Oakland on 27 March 1954. (Doug Olson) Below, VC-35 AD-5N BuNo 132485 at NAS Oakland, CA, on 15 May 1954. (Doug Olson)

receipt of the squadron's first two A-7A Corsair IIs to fulfill its new mission as the A-7 RAG. The Corsairs quickly replaced the Skyraiders until finally the squadron hack, an A-1E, was retired in October 1968.

Above, VA(AW)-35 BuNo 132498 in flight near San Diego on 2 March 1957. (Paul Minert collection) At left, VA(AW)-35 AD-5N BuNo 135047 taxis after trapping aboard the USS Essex (CVA-9) in October 1956. (National Archives) Below, VA(AW)-35 AD-5N BuNo 132615 in storage at Davis-Monthan AFB. VA(AW)-35 changed their tail code from "NR" to "VV" on 1 July 1957. (William Swisher) At top right, VA(AW)-35 AD-5Ns BuNos 132610 (VV/823), 132619 (VV/824), and 132552 (VV/825) from the USS Hancock (CVA-19) fly past Mt. Fuji in 1958. The squadron was part of CVG-15. (USN via Katsuhiro Minoura) At right, VA-122 AD-6 BuNo 139665 with da-glo red tail, cowl, and wing stripes. (Paul Minert collection) Below right, VA-122 AD-6 BuNo 135232 in flight with da-glo red/orange engine cowl, tail and drop tank stripes. (Harry Gann)

Above, VA-122 AD-6 BuNo 139645 at MCAAS Yuma, AZ, on 3 December 1959. (William Swisher)

VA(AW)-35 DETS

DET G   VAN 50

CVA-16   1957

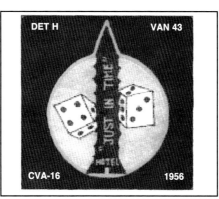

DET H   VAN 43

CVA-16   1956

VAAW-35 FOXTROT VAN 48

USS HORNET CVA 12

CVA-12   1957

CVA-19   1958

CVA-10   1957

DET A   VAN 59

CVA-41   1958-59

At left, VA-122 A-1H BuNo 134622 at Davis-Monthan. (Paul Minert collection) Below left, VA-122 A-1E BuNo 132437 with large yellow tail stripe outlined in black. (Paul Minert collection)

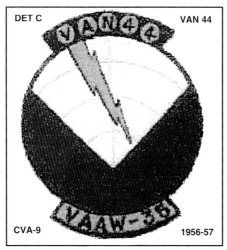

DET C   VAN 44

CVA-9   1956-57

Below, VA-122 AD-5 BuNo 133865 at Lemoore in September 1962 had daglo red tail and wing markings. (William Swisher)

129

THE FIRST ATTACK SQUADRON THIRTY-FIVE, VA-35

Attack Squadron Thirty-Five (VA-35) was first established as Torpedo/Bombing Squadron Two (VT-2) on 6 July 1925 aboard the USS Aroostook (CM-2) at NAS Pearl Harbor, TH. On 1 July 1927, VT-2 was redesignated VT-2B at NAS San Diego where it operated T2D-1s, T3M-2s, TB-1s, T4M-1s, TG-1s and TG-2s before being redesignated Torpedo Squadron Three (VT-3) on 1 July 1937. In October 1937, the squadron re-equipped with the Douglas TBD-1 Devastator which was replaced with TBF/TBMs in July 1942. VT-3 was redesignated Attack Squadron Four A (VA-4A) on 15 November 1946 while still flying a mixed bag of TBMs. On 7 August 1948, VA-4A was redesignated Attack Squadron Thirty-Five (VA-35) while aboard the USS Kearsarge (CV-33).

In April 1949, while under the command of LCDR Norman D. Johnson, VA-35 received one Douglas AD-1 Skyraider which operated alongside seventeen TBM-3Es and one SNJ-5. Sixteen AD-2s were acquired in May 1949 which replaced the AD-1 and all but one TBM-3E. LCDR Roy P. Gee relieved LCDR Johnson on 24 August 1949 at NAS Quonset Pt., RI, before VA-35 was disestablished on 7 November 1949.

Below, VA-35 AD-2s taxiing while folding wings. (SDAM)

FIGHTER SQUADRON FORTY - TWO, VF-42 "GREEN PAWNS"
ATTACK SQUADRON FORTY - TWO, VA-42 "GREEN PAWNS"

Fighter Squadron Forty-Two (VF-42) was established at NAAS Oceana, VA, on 1 September 1950 as part of CVG-6 with F4U-4s. In September 1953, while under the command of LCDR Shelley B. Pittman, VF-42 began receiving AD-4 Skyraiders. CDR Pittman was relieved by LCDR Leroy P. Smith on 8 October 1953 and VF-42 was redesignated Attack Squadron Forty-Two (VA-42) on 30 November 1954. ATG-181 and VA-42 joined the USS Randolph (CVA-15) with AD-6s and AD-4Ns for a Mediterranean deployment which ended on 18 June 1955.

With CDR L. W. Squires at the helm, VA-42 deployed aboard the Navy's first super carrier, the USS Forrestal (CVA-59) during its shake-down cruise in January through April 1956. ATG-181 deployed aboard the USS Bennington (CVA-20) to the Western Pacific from 12 October 1956 until 22 May 1957. The

squadron's final deployment was aboard the USS Intrepid (CVA-11) from 9 June through 8 August 1958 to the North Atlantic with CDR Clifton R. Largess, Jr. as CO.

On 24 October 1958, while under the command of CDR Robert Linwick, Jr., VA-42 was tasked to train fleet replacement pilots in the AD Skyraider. The syllabus included day and night carri-

Above, VA-42 AD-6s staging for launch from the Forrestal during its shakedown cruise in January 1956. (USN) Below, VA-42 AD-6 BuNo 135357 misses the wires and goes off the angle on the Forrestal to try a second pass off Guantanamo Bay in March 1956. (National Archives)

Above, eleven VA-42 AD-6s run-up prior to a deck launch from the USS Forrestal (CVA-59) on 14 March 1956 while attached to ATG-181. (USN) Below, VA-42's only gull grey and white AD-6 used on the Forrestal shakedown cruise was BuNo 137545 taxiing with four other all-blue AD-6s. All aircraft had orange fin tips. The blue birds also had white rudder edges and horizontal tail tips and a small squadron insignia on the upper tail near the leading edge. (USN)

er qualifications, all-weather tactics, special weapons tactics, conventional weapons (bomb, rocket and guns) tactics, and low level navigation flights. The first Skyraider replacement pilot was graduated on 19 February 1959 and on 9 March the T-28B was received for use as an instrument trainer with VA-42 and one to two AD-5s were maintained for the same purpose. On 10 November 1962, three VA-42 pilots conducted the first trans-Atlantic Skyraider flight when their three A-1Hs left Argentia, Newfoundland to Rota, Spain, via Lajes, Azores.

On 1 February 1963, VA-42 received the first fleet deliveries of the A-6A Intruder as the squadron

At top, CVA-11 assigned VA-42 AD-6 BuNo 137537 refuels a VFP-62 F9F-8P photo Cougar in August 1958. Fin stripes were orange and black. (USN) Above, VA-42 AD-6 BuNo 137507 shed its engine on 28 September 1957 on CVA-20 while assigned to ATG-181, tail code "AM". (USN) Below, VA-42 AD-6 BuNo 139797 in flight in 1959 with its Replacement Air Group "AD" tail code. (Paul Minert col.)

became responsible to train A-6 pilots and Bombardier/Navigators as well as maintenance personnel. The first A-6 class, 1-63, convened on 3 September 1963 and the squadron's last A-1H, BuNo 135324, was transferred out on 8 September 1963.

However, VA-42 retained one A-1E and two T-28Bs for use in instrument training for Skyraider pilots. The last T-28B was gone by 12 March 1964, with the squadron's A-1E being transferred out on 14 March 1964.

Above, VA-42 AD-5 BuNo 132443. (USN) Below, VA-42 AD-6 BuNo 137624 in flight near NAS Oceana, VA. Fin stripes were green and black. Thin cowl stripe was green. (USN via Jim Sullivan)

THE SECOND ATTACK SQUADRON FORTY - FOUR, VA-44 "HORNETS"

Fighter Squadron Forty-Four (VF-44) was established on 1 September 1950 at NAS Jacksonville, FL, with F4U-5s. After two deployments to the Mediterranean aboard the USS Coral Sea (CVB-43) in 1951 and 1952, the squadron deployed to Korea aboard the USS Lake Champlain (CVA-39) as part of CVG-4 in 1953, then transferred to the USS Boxer (CVA-21)

and ATG-1 for three months before returning to "The Champ" and CVG-4. After this deployment, the squadron relinquished its Corsairs for the F2H-2 Banshee. The F2H-2s were used on the USS Intrepid (CVA-11) in 1955. Then on 1 January 1956, VF-44 was redesignated Attack Squadron Forty-Four (VA-44) at Jacksonville where they received F9F-8 Cougars in April 1956. In late 1957, a detachment of Cougars were deployed aboard the USS Wasp (CVS-18) to protect the sub hunters and the ship.

On 1 June 1958, VA-44's mission was changed to that of the fleet replacement training squadron for the East Coast Skyhawk squadrons. They were tasked at training both pilots and maintenance personnel. As such, the squadron received F9F-8Ts, A4D-1s, TV-2s, T-28Bs, and A4D-2s in 1958. On 6 June 1958, Fleet All-Weather Training Unit

Detachment Alpha was disestablished and its instrument training mission, aircraft and personnel were transferred to VA-44.

In January 1959, the squadron began to take over the replacement training of Skyraider pilots and maintenance personnel and took delivery of two AD-5s and six AD-6s to complete this mission. This mission was suspended on 15 February 1963 with the squadron concentrating on Skyhawk training while flying A-4B/Cs and TF-9Js. When withdrawn, the squadron had two A-1Es and 15 A-1Hs on hand. VA-44 was disestablished on 1 May 1970.

Below, VA-44 AD-6 BuNo 139765 at NAS Norfolk in 1962. Fin tip was orange. (Norm Taylor collection)

Above, four VA-45 AD-2s over the Pinecastle, FL, target range on 13 February 1952. (USN) Below, VA-45 AD-6 (F/504) BuNo 135315 waves-off from the USS Intrepid (CVA-11) during carrier qualifications in February 1955. (Swisher collection)

In response to the Korean War, the second Attack Squadron Forty-Five (VA-45) was established on 1 September 1950 at NAS Jacksonville, FL. Seventeen days later, the squadron transferred to NAAS Cecil Field, FL, where it received its first two AD-2 Skyraiders before the end of the month. Sixteen AD-2s were on-hand on 31 October, and the squadron's full complement of eighteen AD-2s was acquired in November. LCDR George O. Wood took his squadron aboard the USS Oriskany (CV-34) on its first deployment to the Mediterranean as part of CVG-4 from 15 May to 4 October 1951.

In January 1952, VA-45 began to transition to the AD-4/4L in preparation for the squadron's second deployment. They finished the month with two AD-4Ls and sixteen AD-2s. At the end of February, they had twelve AD-4s and four AD-4Ls. On 11 April, LCDR Richard H. Mills replaced LCDR Wood as CO. Eight days later, VA-45 deployed to the Med aboard the USS Coral Sea (CVB-43) returning on 12 October 1952. Upon returning to Florida, the squadron transferred back to Jacksonville where LCDR William F. Krantz took command.

In preparation for VA-45's third deployment, the squadron upgraded its AD-4/4Ls to AD-4Bs for the war in

Korea. The unit took sixteen AD-4Bs aboard the USS Lake Champlain (CVA-39) on 26 April 1953 for a combat cruise to Korea, returning on 4 December. VA-45's first combat missions were flown on 13 June 1953. Only one pilot, LTJG Donald Brewer, was lost in Korea; he was shot down west of the Punchbowl by 37mm flak. For its service, the squadron was awarded the Korean Presidential Unit Citation, the Korean Service Medal, and the United Nations Service Medal.

The AD-6 began replacing the AD-4Bs in June 1954 with six on-hand at the end of the month. In August, a full complement of sixteen AD-6s was achieved and in November 1954 CDR Daniel W. Wildfong relieved LCDR Krantz. The

squadron returned to the Med for its fourth deployment aboard the USS Intrepid (CVA-11) from 28 May to 22 November 1955.

In August 1956, CDR Glendon Goodwin, CO of VA-45 since December 1955, made the first propeller aircraft landing aboard the USS Saratoga in his AD-6 during CVA-60's shakedown cruise. The squadron's last deployment was to the Med aboard the USS Randolph (CVA-15)

from 1 July 1957 to 24 February 1958. On return to Jacksonville, the squadron was disestablished on 1 March 1958.

Above, VA-45 AD-6 BuNo 135284 taxis on CVA-11 on 22 February 1955. Trim was white. (National Archives) Below, VA-45 AD-4B BuNo 132256 takes the barrier aboard the USS Lake Chaplain, CVA-39, on 9 May 1953. The pilot was LTJG L.E. Brumbach. (National Archives)

Above, VA-45 AD-6 BuNo 134538 deck launching from the USS Saratoga (CVA-60) during her shakedown cruise on 3 September 1956. Fin tip was orange. (USN) Below, the VA-45 team sent to the second annual weapons meet at El Centro, CA, in April 1957 break for their landing. From bottom to top; BuNo 137542 (F/402), 137551 (F/404), 137503 (F/403), and 137??? (F/401). (National Archives)

THE THIRD ATTACK SQUADRON FORTY - FIVE, VA-45 "BLACKBIRDS"

The third Attack Squadron Forty-Five (VA-45) at NAS Jacksonville, FL, was established on 15 February 1963 as the Atlantic Fleet A-1 Skyraider replacement training squadron with the personnel and aircraft from VA-44 (see page 135) that had been performing that function. They inherited fifteen A-1Hs, two A-1Es, and four T-28Bs and took over the previous VA-45's insignia and traditions (see pages 137-138).

In April 1964, the squadron's mission changed from the A-1 RAG to the Atlantic Fleet instrument training squadron. With this change in mis-

sion, the Skyraiders were replaced with Grumman TF-9J Cougars. They continued to operate the T-28Bs alongside the TF-9Js and moved to NAS Cecil Field, FL. The squadron started phasing out their TF-9Js in March 1967 when their first TA-4Fs arrived. The T-28Bs left in 1969 and

---

Below, VA-45 A-1H on 15 April 1963 loaded with eighteen bombs and twelve rocket pods. Tail code was "AD". (USN) Bottom, bombed-up VA-45 A-1E BuNo 133913. (USN)

Above, VA-45 A-1E BuNo 132466 in July 1965. (Dave Menard) At left, VA-45 A-1E BuNo 132466. (Ginter collection) Below, VA-45 A-1H BuNo 137553 during carrier qualifications in 1963. Fin tip was black. (Norm Taylor collection)

the squadron was redesignated VF-45 on 7 February 1985. VF-45 was disestablished in March 1996.

THE FIRST ATTACK SQUADRON FIFTY - FOUR, VA-54

Scouting Squadron Two B (VS-2B) was established on 3 July 1928 at NAS San Diego, CA, and flew UOs, FU-1s, and O2U-1/2s into 1931. The O3U-2 was acquired in December 1931 and in 1932 the SU-1/2/3 were received. This was followed by the SBU-1 in January 1936 with the squadron being redesignated VS-3 on 1 July 1937. In August 1937, the SBC-3 bi-plane Helldiver was received followed in March 1941 by the SBC-4. In August 1941, the Douglas SBD-3 replaced the early Helldivers and were supplemented with SBD-4s in January 1943. On 1 March 1943, VS-3 became Bombing Squadron Four (VB-4) and the Curtiss SB2C Helldiver was added. Then, on 15 July 1943, VB-4 was redesignated VB-5. In 1944, the squadron re-equipped with SB2C-3/4/4E and SBW-3 Helldivers and finished the war with them. The SB2C-5 was acquired in March 1946 and was utilized until May 1949. VB-5 was redesignated Attack Squadron Five A (VA-5A) on 15 November 1946 at San Diego where VA-5A was redesignated VA-54 on 16 August 1948. LCDR D. K. English was the commanding officer when VA-54 began receiving the AD-1 Skyraider in April 1949. VA-54 was the last fleet SB2C-5 squadron at the time.

VA-54 finished April 1949 with six AD-1s and 15 SB2C-5s on-hand. In May, there were thirteen AD-1s and three SB2C-5s and in June through September the squadron had a full complement of fifteen AD-1s. In October, the squadron was scheduled to upgrade to the AD-4 Skyraider. It received thirteen AD-4s that month and retained sixteen AD-1s, but all aircraft were transferred out in November and the squadron was disestablished on 1 December 1949 at NAS San Diego. For the mere six months the squadron operated the Skyraider, it was assigned to CVG-5.

Below, VA-54 AD-1 (S-407) buzzes the tower at NAS San Diego, CA, while armed with twelve dummy HVARs, a 500 lb bomb, a 750 lb bomb and an APS-4 radar pod on 2 June 1949. (USN)

Above, VA-54 AD-4 (S/401) weapons display at NAS San Diego, CA, on 7 November 1949. Fuel load was represented by seven 55-gallon drums, one 13" torpedo for the centerline, two 1,000 lb bombs for the inboard wing pylons, twelve HVAR rockets for the outer wing stations, and two 20mm cannons with ammo. (National Archives) Below, VA-54 AD-1 Skyraider (S/407) in June 1949 loaded out with yellow practice ordnance. Pilot was LTJG O'Neill. Rocket noses were red. Fin tip was yellow. (USN)

Bomber Fighter Squadron One Hundred Fifty-Three (VBF-153) was established on 26 March 1945 at NAS Wildwood. The squadron was initially equipped with F4U Corsairs and received Grumman F6F-5 Hellcats in September 1945. VBF-153 was redesignated Fighter Squadron Sixteen A (VF-16A) on 15 November 1946. The squadron made one deployment with the Hellcat before receiving F8F-1 Bearcats on 21 October 1947. VF-16A was redesignated VF-152 on 15 July 1948 at NAS Alameda, CA. The F8F-2 was received in May 1949 and the F4U-4B in December 1949. VF-152 had sixteen F8F-2s and sixteen F4U-4Bs in December 1949 and by the end of January 1950 only the Corsairs remained. On 15 February 1950, VF-152 was redesignated VF-54. The squadron deployed to Korea aboard the USS Valley Forge (CV-45) with the Corsair from May through December 1950.

In March 1951, VF-54 received six AD-1s at NAS San Diego, CA, while sixteen F4U-4s were aboard CV-45 for training. By the end of April 1951, all the Corsairs were gone and

At right, four VF-54 AD-4s in flight on 23 April 1950 with BuNo 123814 at top. (National Archives)

through 30 January 1952, he ditched his disabled Skyraider into Wonsan Harbor. After the fourth ditching, the Admiral restricted him to flying anti-submarine patrols. A sign was posted in the ready room, "Use caution when ditching damaged airplanes in Wonson Harbor. Don't hit Commander Gray". A total of nine aircraft were lost due to enemy action and five to operational failures during combat missions.

A high value raid without loss of aircraft or crews was conducted on 29 October 1951. The target was a high level meeting at the headquarters of the Chinese Communist Party near Kpsan. LCDR Gray led the first division with LTJG Shugart, ENS Aillaud, and ENS Masson. The second division was comprised of LT Evans,

the squadron had nine AD-4Ls, two AD-4s and five AD-1s. When LCDR Paul N. Gray took VF-54 to Korea on 28 June 1951, it was equipped with thirteen AD-4Ls, five AD-4s, one AD-3N, one AD-3Q, one AD-4Q, and three AD-4Ws. CVG-5 sailed aboard the USS Essex (CV-9) and returned to San Diego on 25 March 1952. On 26 October 1951, LTJG William Burgess ditched off Hungnam Harbor and was rescued by the USS Conway

after being hit by two 88mm shells. For its Korean War deployment, VF-54 received the Navy Unit Commendation and the Korean Presidential Unit Commendation.

During the Essex deployment, LCDR Gray became known as the "Bald Eagle of the Essex" in recognition of being the most rescued naval aviator of the Korean War. Four times, from 7 September 1951

Above left, VF-54 AD-4N BuNo 125751 after an emergency landing in Korea. Fin tip was yellow. (Paul Minert collection) Below, VF-54 AD-4 formation on 23 April 1950. (National Archives)

LTJG Gollner, ENS Strickland, and ENS Kelly. Each aircraft carried a proximity fuzed 1,000lb bomb, an impact fuzed 1,000lb bomb, a napalm bomb and eight 250lb bombs. The proximity fuzed bombs were dropped first followed by the standard 1,000lb bomb on the next pass. Then four napalm bombs were dropped during a straffing run followed by another run in which the last four napalm bombs were dropped. On the last pass the 250 pounders were let loose and the communist compound was completely destroyed. Post strike

Above, VF-54 AD-3 BuNo 122737 with flak damaged wing at K-18 Korea on 23 January 1952. Fin tip was yellow. (USN) Below, VF-54 AD-4NA BuNo 125750 taxis at Yodo Island, Wonson Harbor, off North Korea in May 1953. (NMNA)

Above, door on one of four VF-54 AD-4NAs modified in-house with four passenger seats in the aft cabin and dubbed VR-54 SCROD Airlines. (USN) Above left, VF-54 AD-4 BuNo 123928 noses-over on landing aboard CV-9 in May 1951. Note AD in the groove has begun its "go around". (via Tailhook) At left, VF-54 AD-4NA taxis on 4 October 1952. (USN) Bottom, VF-54 AD-2 loaded with bombs and napalm deck launches from the USS Essex (CV-9) on 1 March 1952. (National Archives)

intelligence reported 509 top level Communists and their records destroyed. The strike team was called "The Butchers of Kapsan" by the Chinese and a price was put on their heads.

On 19 May 1952, VF-54 was transferred to NAS Miramar, CA, where LCDR Henry Suerstedt, JR. assumed command on 27 June 1952. After training was completed, VF-54 deployed again to Korea on 20 November 1952. They fielded eight AD-4s and eight AD-4NAs aboard the USS Valley Forge (CVA-45). The deployment was concluded on 25 June 1953 and the squadron received a second Navy Unit Commendation and Korean Presidential Unit Commendation for its actions while flying the Skyraider in combat over Korea.

Prior to the deployment, VF-54 installed four seats in the aft compartment of four of their AD-4NAs. The

planes were used to transport personnel from ship to shore or more usually in Korea from shore to ship as they often picked up pilots who had made emergency landings ashore. The crew doors were marked VR-54 SCROD Airlines. SCROD stood for a baby COD (carrier on-board delivery).

Once back at Miramar, LCDR Christian Fink took command on 21 July 1953. On 12 March to 6 November 1954, VF-54 took its AD-4 on a WestPac deployment aboard the USS Philippine Sea (CVA-47). On 26 July, VF-54 Skyraiders and one VC-3 F4U shot down two Chinese LA-7 fighters who had previously downed a British Cathay Pacific DC-4 airliner on 22 July. CVG-5 was conducting a

Above, VF-54 AD-4 deck launches from the USS Valley Forge (CV-45) in 1952-53. Fin cap and prop hub were yellow. (Harry Gann) Below, VF-54 AD-6 BuNo 135366 at NAS Alameda, CA, on 5 August 1956. Rudder was gull grey too. (William T. Larkins)

SAR mission and was looking for survivors, rafts and debris when attacked by the LA-7s. The encounter became known as the "Hainan Turkey Shoot". LT Roy M. Tatham and ENS Richard R. Crooks of VF-54 were credited with splashing one LA-7 and VF-54's XO LCDR Paul Wahlstrom, LTJGs John Damian, Richard Ribble, and John Rochford shared credit for the

second LA-7 with VC-3's LCDR E.B. Salsig.

As VF-54, one more deployment was made from 29 October 1955 to 17 May 1956. It was another WestPac cruise, this time with AD-6s aboard the USS Kearsarge (CVA-33) while under the command of CDR Frank M. McLinn. After the squadron's return to CONUS, VF-54 was redesignated Attack Squadron Fifty-Four (VA-54) on 15 June 1956

and CDR William A. Lewiston relieved CDR McLinn.

One deployment was made as VA-54 with AD-6/7s. A WestPac cruise aboard the USS Bon Homme Richard (CVA-31) from 12 July to 9 December 1957. In September 1957, the squadron operated off Taiwan to quell tensions caused by a buildup of Chinese Communist forces on the mainland across the strait. Upon its return to Miramar, VA-54 saw CDR

Above, VA-54 AD-7 trapping on CVA-31 in 1957. (USN) Below, VA-54 AD-7 BuNo 142069 on the Bon Homme Richard (CVA-31) in late 1957. Drop tanks, cowl, and fin flashes were orange. (USN)

Emmit W. Blackburn assume command on 16 December 1957 and the receipt of F9F-8Bs that same month. VA-54 was disestablished on 1 April 1958.

ATTACK SQUADRON FIFTY - FIVE, VA-55 "TORPCATS"/"WARHORSES"

VA-55 was first established as Torpedo Squadron Five (VT-5) on 15 February 1943 at NAS Norfolk, VA. VT-5 was equipped with TBF-1/1Cs before receiving TBM-3s in September 1944 and TBM-3Es in June 1945. They were flying TBM-3Es and TBM-3Qs on 15 November 1946 when VT-5 was redesignated Attack Squadron Six A (VA-6A). At NAS San Diego, CA., VA-6A was redesignated Attack Squadron Fifty-Five (VA-55) on 16 August 1948 while still flying TBMs.

In June 1949, VA-55 was assigned to CVG-5 and received thirteen AD-1s, all but one being traded out for AD-4s in October 1949. One AD-4Q, BuNo 124047, was added in March 1950 in preparation for its upcoming Korean War deployment aboard the USS Valley Forge (CV-45).

VA-55 became the first Skyraider

TORPCATS 1946-1955

squadron to strike the enemy when on 3 July 1950 they struck Pyongyang and Onjong-Ni airfields. Led by their CO, LCDR N.D. Hobson, they carried two 500lb and six 100lb bombs destroying and damaging hangars, barracks, aircraft, and a fuel farm. Four AD-4s were hit but returned to fight another day. The next day the squadron struck again. The North Koreans lost a bridge span and twelve locomotives and once

again the unit's ADs took hits, but all pilots returned safely even though AD-4, BuNo 123804, crashed on landing taking out two Corsairs and damaging three ADs. And so it went for the rest of the war, with interdiction

---

Below, VA-55 AD-4 loaded with bombs and rockets runs-up aboard the USS Valley Forge (CV-45) on 8 July 1950. (National Archives)

and close air support being the Skyraiders key missions.

On 18 July 1950, eleven VA-55 Skyraiders loaded with one 1,000lb bomb, two 500lb bombs, and four rockets along with 10 rocket armed VF-53

Corsairs struck the Wonsan oil refinery. The strike totally obliterated the facility and burned some 12,000 tons of refined petroleum providing the communists a major setback,

A tragic loss occurred on 22 July

Above, VA-55 AD-1 BuNo 09333 at NAS Glenview, IL, in 1949. (William Swisher) Below, VA-55 AD-4Q BuNo 124047 in December 1950 after the squadron's return to San Diego from its first war deployment. Fin tips were green. (NMNA)

1950 when the pilot of AD-4 BuNo 123844 spun in to avoid straffing a truckload of civilians. On 7 August, four VA-55 AD-4s scored three direct hits on the Kochang bridge and damaged the Koksong bridge. Another strike on the 7th with six AD-4s hit the Seoul highway bridge with little effect, so a second strike hit the bridge, but it was still standing. Finally, the 8-inch guns on the USS Helena finished it off. On 25 October 1950, BuNo 123808 was hit by small arms fire and forced to ditch at sea. Two pilots brought back damaged ADs after being fragged by their own bombs.

The squadron returned to Korea by ship and air transport to join CVG-19X aboard the USS Princeton (CV-37) from 16 May to 29 August 1951. LCDR A.L. Maltby Jr. was in command of sixteen AD-4s, two AD-4Qs, two AD-4Ws and one AD-4N.

Three aircraft were lost in June.

During a CAP flight, AD-4 BuNo 123931 ditched after an engine failure on 6 June. Then, AD-4 BuNo 123932 ran afoul while landing on CV-37 on 21 June. And finally, AD-4 BuNo 123855 was downed by small arms fire on 28 June.

Above, VA-55 AD-4B BuNo 123952 after trapping aboard the USS Essex (CVA-9) in 1954 while assigned to ATG-2. Lightning bolt on the tail was white. (USN) Below, VA-55 relieved VA-195 aboard CV-37 on 31 May 1951. VA-55's AD-4 BuNo 123821 was armed with three 2,000lb bombs, six 100lb bombs, and six 5" HVAR rockets prior to a dam busting mission over Korea. (USN)

151

VA-55's third war cruise was aboard the USS Essex (CVA-9) from 16 June 1952 to 6 February 1953 as part of ATG-2 with CDR L.W. Chick commanding.

In 1952, five squadron pilots went down due to enemy action and all were recovered. LT John Page had his prop blown off while on a close air-support mission and crash landed near the front behind UN lines. LT Jim Norton ditched at sea and was picked up by a destroyer after being hit by flak on an armed recon mission. The third pilot, LT Tom Davenport, also ditched at sea and was picked up by a destroyer after being hit by AA. The next two pilots downed encountered much greater difficulty in returning to Essex. ENS Peter Moriarty was hit on a rescue mission and was forced to bail out over North Korea. When he landed, there were two soldiers waiting for him and his Keystone Cop escape began. At some point one of the soldiers emptied his revolver on

Above, VA-55 AD-4 burns on the flight deck of CVA-9 on 17 March 1954. (USN) Below, VA-55 AD-6 BuNo 139651 while assigned to ATG-2 in 1956. Wing tip and rudder trim was green and white checkerboards. Fin tip was green. (Paul Minert collection) Bottom, VA-55 AD-6 BuNo 139650 deck launching from CVA-19 as part of ATG-2 carrier qualifications in 1957. (USN)

Moriarty who was standing five feet away. Moriarty ran as all the bullets missed. He ran into an adjacent field because a UN helicopter was coming in to rescue him. As the chopper set down he jumped in the door amist rifle fire from approaching troops. The chopper made it back to base with half-a-dozen bullet holes in it.

While attacking the Kojo hydro-electric plant, LTJG John Lavra was hit by flak and his AD burst into flames. He found himself spinning toward the ground from 4,000 ft. Finally, at about 1,000 ft he got out. Despite his burns he was able to evade his pursuers until rescued by helicopter.

The squadron's first post-war cruise was aboard the USS Essex (CVA-9) from 1 December 1953 to 12 July 1954 with LCDR R.J. Thompson commanding.

VA-55 took their AD-6s aboard the USS Philippine Sea (CVA-47) from 1 April to 23 November 1955. Ports of call were Pearl Harbor, Yokosuka, Hong Kong, Keeling, and Formosa. As part of CVG-2, joint operations with Chinese Nationalist Forces off Formosa was conducted.

The squadron's last deployment was aboard the USS Hancock (CVA-19) as part of ATG-2 from 6 April to 18 September 1957 under the command of CDR Billy D. Holder.

After returning home to Miramar, the squadron started drawing down its complement of AD-6/7s in preparation for their transition to the FJ-4B. The first nine Fury Bravos were received in November and by 31 December 1957 all the Skyraiders were gone.

Above, VA-55 XO LCDR Farrell poses with his mount (NB-502) on CVA-19 in 1957 as part of ATG-2. (USN) Below, VA-55 AD-6 BuNo 139652 taxiing forward aboard CVA-19 in 1957 while assigned to ATG-2. (USN) Bottom, six VA-55 AD-6s in flight from CVA-19 in 1957. (USN)

THE FIRST ATTACK SQUADRON SEVENTY - FIVE, VA-75

The first VA-75 was originally established as Torpedo Squadron Eighteen (VT-18) on 20 July 1943 at NAS Alameda, CA. Throughout WWII and most of the remainder of the 40s the squadron flew various versions of the Grumman Avenger. On 15 November 1946, VT-18 was redesignated Attack Squadron Eight A (VA-8A) while home-based at NAS Quonset Point, RI. VA-8A was redesignated Attack Squadron Seventy-Five (VA-75) on 27 July 1948.

While assigned to CVG-7, half the squadron deployed with its TBMs aboard the USS Philippine Sea (CV-47) from 4 January to 23 May 1949. The other half remained at Quonset Point for receipt and transition to the AD-3 Skyraider. The first aircraft was acquired on 18 April and by 31 May VA-75 had sixteen AD-3s and one AD-2Q on-hand. CDR Morris R. Doughty was in command for the receipt of the Skyraiders and for the squadron's disestablishment on 30 November 1949.

Below, VA-75 AD-3 taxis forward with wing braces in place in 1949. Fin tip and prop hub were green. (USN)

ATTACK SQUADRON SEVENTY - FOUR, VA-74 "SUNDAY PUNCHERS"
THE SECOND ATTACK SQUADRON SEVENTY - FIVE, VA-75

The second VA-75 was originally established as Bombing Squadron Eighteen (VB-18) on 20 July 1943 at NAS Alameda, CA. VB-18 was originally equipped with SBD-5s before transitioning to Curtiss Helldivers in March 1944. On 15 November 1946, VB-18 was redesignated Attack Squadron Seven A (VA-7A) while home-based at NAS Quonset Point, RI. In April 1948, the Corsair replaced the Helldiver and on 27 July 1948 VA-7A was redesignated Attack Squadron Seventy-Four (VA-74).

The squadron was commanded by LCDR William B. Morton when, in

May 1949, the squadron's first three AD-3s were acquired. By 30 June

1949, VA-74 had sixteen AD-3s and three F4U-4s which were transferred

out in July. On 15 February 1950, while under the command of LCDR Nils R. Larson, VA-74 was redesignated Attack Squadron Seventy-Five (VA-75) and in June, the squadron's first nine AD-4s were received. VA-75's first AD deployment began the following month on 10 July as part of CVG-7 aboard the USS Midway (CVB-41). The cruise to the Mediterranean with eighteen AD-4s was concluded on 10 November 1950.

The squadron's second AD cruise was aboard the USS Bon Homme Richard (CVA-31) from 20 May 1952 to 8 January 1953 with LCDR H.K. Evans commanding. VA-75 flew its first combat missions over Korea on 23-24 June 1952 when North Korean hydro-electric complexes were struck. Follow-up attacks on the plants were made on 3 July. This was followed by 4-days of strikes against transportation, troop barracks, and supply warehouses in the Wonsan Valley. On 9 July, AD-4 BuNo 128969, was ditched after take-off. Then on 11 July the squadron accompanied elements from CV-37, USAF, USMC, RAAF, and RN on a strike against Pyongyang. The power plants were struck again on 4 August with the transformer stations, switch houses and supply houses being targeted. VA-75 attacked Hamhung, Pukchong, and the Sindok lead and zink mines on 9 August and destroyed five coastal defense guns on the 11th. Early September saw strikes conducted in behalf of the US

First Corps and the Second ROK followed on 10 September with a return strike to the Kyosen power plants. The lead and zinc mines were hit again on the 12th and on the 24th the Sungjibaegam electro-metallurgical plant was destroyed. A four day operation called "Kojo Amphibious Feint" was conducted on 12-15 October during which AD-4, BuNo 127875, was

Above, VA-75 AD-4 BuNo 123904 lifts off from the USS Tarawa (CV-40) for a training strike near Guantanamo Bay, Cuba, on 3 May 1951. (National Archives) Below, destroyed Kyosen No. 4 generator house struck on 25 June 1952. (USAF) Bottom, VA-75 AD-4 from CV-31 deck launches on the Thanksgiving Day raid over North Korea on 27 November 1952. Fin tip was green. (National Archives)

ditched on the 15th due to AAA fire. Additional strikes against the hydro-electric plants in November put them out of commissioned for the rest of the war. The CO, LCDR Evans, was killed-in-action on 5 December after his AD-4, BuNo 128965, went down due to AAA fire. LCDR W.M. Harnish assumed command as acting CO for the remainder of the deployment. A total of 1,216 combat missions were flown over Korea.

For the squadron's third AD-4 deployment they boarded the USS Bennington (CVA-20). The North Atlantic/Med cruise was from 16 September 1953 to 21 February 1954 with eight AD-4Bs, one AD-4L, and five AD-4s. On 23 September 1953, during Operation Mariner CVA-18, CVA-20, and HMCS Magnificent were assigned to "Blue Force". Blue Force launched with 52-aircraft about noon. There were thirteen AD-4B/Ls and one AD-4Q from VA-75 and an additional thirteen AD-4s from Wasp's VA-175. During the strike force run in, the weather deteriorated rapidly into a solid undercast and at

1420 hours ADM Goodwin aboard CVA-20 ordered a recall. Fourteen Corsairs returned of which ten trapped before fog completely enveloped the flight deck. When the VA-75 Skyraiders returned, they dropped to 300 feet in a fruitless effort to find VFR conditions for landing. They then entered a 3,000 ft holding pattern and a couple of pilots were tasked to attempt radar vector approaches. These also failed due to zero visibility and the ship's loss of radar to sea return. At 1705 hours, the USS Redfin (SS-272) reported a clearing 110 miles west with a 1,000 ft ceiling. All aircraft were ordered to proceed to Redfin to ditch along-side. The flight to Redfin required the air-craft to pass over TF 218 on their way west and VA-75's CO, CDR Benjamin Preston, saw a small hole in the undercast. The fog over the Task Force began to quickly weaken and soon surface visibility became a ceiling of 300 ft and two miles. The three carriers turned into the wind in a line-abreast formation. The approaches were still made in great difficulty as night was falling and the ceiling and

visibility changed from minute-to-minute. The last aircraft touched down at 1828 hours and miraculously none were lost. One VA-75 AD was recovered aboard Magnificent and three more aboard Wasp.

In August 1954, VA-75 received its first two AD-6s and all versions of the AD-4s were gone by the end of November. The AD-6/A-1Hs would deploy six times with VA-75. The first aboard the USS Hornet (CVA-12) to the Western Pacific from 4 May to 10 December 1955. During the cruise, they participated in "Operation Passage to Freedom", the evacuation of citizens from North Vietnam to South Vietnam.

The second AD-6 cruise was from 3 September to 22 October 1957

Below, VA-75 AD-4 BuNo 123934 being hoisted aboard the USS Iowa from a barge off of Wonsan, Korea, in September 1952. (National Archives via Jim Sullivan)

Above, VA-75 AD-4 piloted by LTJG Jim Elster runs-up for launch from the Canadian carrier Magnificent on 25 September 1953 at the conclusion of Operation "Mariner Miracle". (NMNA) Below, VA-75 AD-4B BuNo 132286 in the weeds during a snow storm at NAF Charlestown, RI, on 7 January 1953. The aircraft's back is broken and the engine has separated from the airframe. Pilot was ENS McCausland. (Larry Webster) Bottom, VA-75 AD-6 BuNo 137613 taxis at NAS Quonset Pt., RI, while folding its wings. (Warren Ship via NMNA)

aboard the USS Saratoga (CVA-60) for its shakedown cruise. During the cruise, VA-75 participated in NATO exercise "Strikeback" with 150 warships from the US, Canada, Netherlands, Norway, and England.

The third AD-6 deployment was aboard the USS Randolph (CVA-15) to the Med from 2 September 1958 to 12 March 1959. Prior to this cruise, the squadron had operated one AD-5 during their CarQual period.

The next three deployments were aboard the USS Independence (CVA-62). These were

Above, VA-75 AD-4B BuNo 132362 in June 1953. Pilot was LT M.A. Thompson. The fin tip was green and wing and horizontal tail tips as well as the rudder trailing edge were white. (Gordon Williams) Below, VA-75 AD-4 (L-501) clears the bow of CVA-20 in 1954. (USN) Bottom, VA-75 AD-6 BuNo 139814 in post-1956 color scheme with green fin tip. (Ginter collection)

from 4 August 1960 to 3 March 1961, 4 August to 19 December 1961, and from 19 April to 27 August 1962.

During the 1960-61 deployment, VA-75 made over 1,400 day and night carrier landings and led Air Group 7 in hours flown. Seventeen of the unit's twenty pilots became "Indy" Centurions on the cruise. It also received the CNO Safety Award for over 15,000 flight hours without an accident. Exercises "Ship-Ring", "Flash Back", "Set-Back", and "Dead Beat" were participated in between six port calls.

VA-75
"SUNDAY PUNCHERS"

Above, VA-75 AD-6 BuNo 135283 on 29 August 1959. Tail, wing, and fuselage trim was green. The fuselage and wing bands were outlined in black. (Tailhook) At right, VA-75 AD-6 BuNo 137438 with twelve 250 lb, two 750 lb, and one 1,000 lb bomb load. (Ginter collection) Bottom, VA-75 AD-6s BuNos 137497 and 139618 aboard CVA-62 on 25 January 1961. (USN)

During the 1961 cruise, the squadron participated in exercises "Checkmate I" and "Checkmate II". In 1962, "Operation Full Swing" was conducted and Indy visited Monaco where Princess Grace and Prince Rainier were entertained.

The squadron's last A-1H Skyraider was retired in September 1963 as the squadron transitioned to the A-6A Intruder.

Above, VA-75 AD-6 BuNo 137498 from CVA-62 on 27 December 1960. (USN) Below, four VA-75 AD-6s BuNos 137498 (AG/503), 135319 (AG/504), 135399 (AG/505), and 139618 (AG/506) in echelon flight, Tail and fuselage trim was green. (Ginter collection)

FIGHTER SQUADRON NINETY - TWO, VF-92 "SILVER KINGS"

Above, blue with yellow wing and fin tips VF-92 AD-4NA after a gear-up landing at NAS Miramar in 1955. (USN) Below, VF-92 AD-4B BuNo 132282 aboard CVA-38 in 1956. (USN) Bottom, VF-92 AD-4B BuNo 132387 on CVA-38 in 1956 while assigned to ATG-3. Fin tip was yellow. (USN)

Fighter Squadron Ninety-Two (VF-92) was established at NAS Alameda, CA, on 26 March 1952 with F4U-4 Corsairs. After a Korean War deployment aboard CV-45, they transitioned to F9F-2 Panthers in 1953. They deployed aboard CVA-47 in 1954 before returning to NAS Miramar and receiving AD-4Bs and AD-4NAs in January 1955. A WestPac AD deployment as part of ATG-3 was conducted aboard the USS Shangri-La (CVA-38) from 5 January to 23 June 1956. Commanded by CDR R.P. Lecklider, VF-92 participated in PACTRAEX 56L and NAVMARLEX during the cruise. In October 1956, the ADs were replaced with F9F-5 Panthers.

ATTACK SQUADRON NINETY - FOUR, VA-94 "TOUGH KITTY"

VA-94 was originally established as Bombing Squadron Ninety-Nine (VB-99) on 1 July 1943 at NAS San Diego, CA. They were redesignated Bombing Squadron Twenty (VB-20) on 15 October and were initially equipped with SBD-5s, but switched to Helldivers in November 1943. On 15 November 1946, VB-20 was redesignated Attack Squadron Nine A (VA-9A) at NAAS Charlestown. VA-9A became Attack Squadron Ninety-Four (VA-94) on 12 August 1948.

Five different versions of Helldivers were flown before the squadron began receiving AD-2s in October 1948 while commanded by LCDR Harlin M. Keister. By February 1949, all the Helldivers were gone and the unit had fifteen AD-2s and one AD-2Q on-hand. VA-94 was disestablished on 30 November 1949 while assigned to CVG-9. Squadron aircraft wore large 400 series aircraft numbers on the after-fuselage and a "D" tail code.

THE FIRST ATTACK SQUADRON NINETY - FIVE, VA-95

The first VA-95 was originally established as Torpedo Squadron Twenty (VT-20) on 15 October 1943 at NAS San Diego, CA, with Grumman Avengers. VT-20 was redesignated Attack Squadron Ten A (VA-10A) on 15 November 1946 while still flying TBMs. On 30 November 1949, VA-10A became Attack Squadron Ninety-Five (VA-95). While under the command of LCDR Charles C. Ainsworth, the AD-1

Skyraider was received in August 1949 while assigned to CVG-9 at NAAS Charlestown. Two months later, the squadron was disestablished on 30 November 1949. VA-95 shared the "D" tail code with VA-94 during this period with VA-95 using large 500 series aircraft numbers on the aft-fuselage sides.

THE SECOND ATTACK SQUADRON NINETY - FIVE, VA-95 "SKYKNIGHTS"

1956

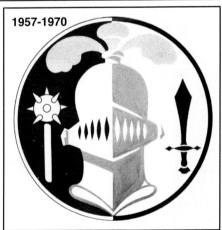

1957-1970

The second Attack Squadron Ninety-Five (VA-95) was established on 26 March 1952 at NAS Alameda, CA, and was first equipped with F6F-5 Hellcats before receiving its first six AD-1s in April 1952. In July, there were two F6F-5s, eight AD-1s, and sixteen AD-4NAs on-hand and by September all the F6F-5s and AD-1s were gone.

The squadron's first CO, LCDR

Samuel B. Berrey, took the unit aboard the USS Philippine Sea (CV-47) for a Korean War combat cruise from 15 December 1952 to 14 August 1953 as part of CVG-9. The first air strikes began on 31 January 1953 and VA-95 concentrated on interdiction and CAS missions. On 15 June, in a maximum effort attack, aircraft from four carriers, CVA-21, CVA-37, CVA-39 and CVA-47, struck in and

around Anchor Hill to support the ROK and US retaking of the hill. During the deployment, LTJG William T. Barron in AD-4 BuNo 123961 was hit by a 37mm shell while attacking two trains. He returned to CV-47 with an 18-inch hole in the right horizontal stabilizer and over two hundred fragment holes in the aft fuselage. On the day CVG-9 was due to rotate home, 27 July 1953, the armistice was

At right, LTJG William Barron points out hole left by 37mm hit on his horizontal stabilizer. Over 200 other holes were counted on his airframe. (USN) Below, VA-95 Skyraider aboard the USS Philippine Sea (CV-47) in 1953. Fin tip was green. (USN)

signed and the war ended, but not before 49 sorties were flown.

In October 1953, AD-6s started replacing the squadron's mixed bag of AD-4s, AD-3s and AD-2Qs. VA-95 redeployed aboard the USS Hornet CVA-12 for a World Cruise from 11 May to 12 December 1954 while under the command of CDR David L. Berrey. Ports-of-call were Lisbon, Naples, then passing through the Suez Canal, Red Sea, and the Indian Ocean before joining the Seventh Fleet in the South China Sea.

On 25 July, while assigned to Task Group 70.2, carrier search planes were attacked by two Communist Chinese fighters which were subsequently shot down. The carrier then visited Hong Kong and Yokosuka before sailing for San Francisco.

During the cruise, in October 1954, CDR Berrey was relieved by CDR John C. Allman.

Under the command of CDR Donald L. Irgens, VA-95 deployed on the USS Oriskany (CVA-34) from 11

Above, VA-95 AD-4N BuNo 127003 in need of repair due to a tail gear failure is flanked by a VF-151 Panther and a VF-194 Skyraider at K-18. (Paul Minert collection) Below, VA-95 AD-3s over the USS Hornet (CVA-12) on 3 October 1954. Fin tips were green. (National Archives)

February to 13 June 1956 to the Western Pacific. Upon return to CONUS, CDR Martin J. Stack took command in September 1956.

CDR Stack took VA-95's eleven AD-7s aboard the USS Ticonderoga (CVA-14) for another WestPac cruise from 16 September 1957 to 25 April 1958. After returning to Alameda, the

Above, armed VA-95 AD-4 from CVA-47 in 1953. (USN) Below, VA-95 Skyraider catapults off the USS Philippine Sea (CVA-47) while at anchor at Ford Island, TH, on 10 January 1953. (National Archives)

At left, VA-95 AD-3 BuNo 122799 on 17 October 1953. (William T. Larkins) Below left, VA-95 AD-7 on 31 January 1956 at NAS Alameda, CA. (USN) Bottom left, VA-95 AD-7 BuNo 142020 in late 1956 with green tail and fuselage trim and a green 9 for CVG-9. (Paul Minert collection) At right, another VA-95 AD-7 taxis at Alameda in late 1956. (Paul Minert collection) Below, VA-95 AD-7 BuNo 142079 had green trim on the tail and fuselage. (Tailhook) Bottom, VA-95 AD-7 BuNo 142010 taxis forward after trapping aboard the USS Ranger (CVA-61) in January 1960. (Hal Andrews collection)

squadron passed through the hands of CDR F.L. Brady (21 March 1958), CDR Rollin E. Gray, Jr. (20 April 1958), CDR Carl Weisse (4 March 1959) before CDR R.R. Renaldi took VA-95 aboard the USS Ranger (CVA-61) from 6 February to 30 August 1960.

A second CVA-61 deployment was made from 11 August 1961 to 8 March 1962 with CDR S.F. Abele at the helm. Ports-of-call were Pearl Harbor, Yokosuka, Sasebo. Kobe, Buckner Bay, Iwakuni, and Hong Kong.

A third Skyraider deployment was made from 9 November 1962 to 14 June 1963 with CDR Abele still in command. Off the Philippines they conducted a simulated war game in conjunction with the Kitty Hawk and the HMS Hermes before reporting to the South China Sea to monitor tensions in Laos.

Above, VA-95 AD-7 BuNo 142048 deck launches from the USS Ranger (CVA-61). (Ginter collection) Below, six VA-95 AD-7s at NAS Barbers Pt., HI, on 21 August 1961 while assigned to the USS Ranger (CVA-61). (USN) Bottom, VA-95 AD-7 BuNo 142038 with a green fin tip and a green number 9 outlined by white. Left wing load-out was two HVARs, two aerial mines, one low drag bomb, a napalm tank and a 150-gallon fuel tank. Fin tip was green. (Paul Minert collection)

In late May, a one day cruise for 150 members of the Japanese-American society was conducted before the ship departed Japan on 3 June 1963.

The squadron's last Skyraider deployment, its 4th aboard Ranger, was from 5 August 1964 to 6 May 1965. It was the unit's only A-1H/J Vietnam War deployment. The squadron, commanded by CDR Dwight DeCamp, arrived on station in November 1964 and commenced Yankee Team operations. In February 1965, the VC attacked the US advisors barracks in Pleiku on the

Above, VA-95 AD-7s BuNos 142074 (NG/512), 142010 (NG/501), and 142060 (NG/507) at Tachikawa AB on 22 May 1960. (Toyokazu Matsuzaki) Below, VA-95 AD-7s at Tachikawa AB on 22 May 1960 while assigned to the USS Ranger (CVA-61). (Toyokazu Matsuzaki) Bottom, VA-95 AD-7 BuNo 142074 traps aboard CVA-61 in 1960. (USN)

7th and the barracks at Qui Nhon on the 10th. Two retaliatory strikes were ordered, dubbed Flaming Dart I and Flaming Dart II. VA-95 took part in both strikes with the Chanh Hoa bar-racks near Dong Hoi being the target on the 11th. On 15 March, the unit participated in the first carrier strike of the Rolling Thunder campaign. The target was an ammunition depot in

Above, four VA-95 AD-7s BuNos 142047 (NG/505), 142020 (NG/503), 142032 (NG/509), and 142056 (NG/506) over the Pacific. (Ginter collection) Below, two VA-95 Skyraiders pass Mt. Fuji in Japan in 1962. (USN)

Phu Qui, where LTJG Charles F. Clydesdale's A-1H, BuNo 135375, was struck by AAA. He subsequently crashed at sea and was killed. On 9 April, they destroyed the Tam Da railway bridge and on 10 April they dropped the Kim Kuong Highway Bridge. Another pilot and aircraft were lost on 11 April during a Steel Tiger armed recon mission. LTJG William E.

Swanson's A-1H, BuNo 135226, was hit by AAA east of Ben Senphan and crashed in the jungle. VA-95 sailed for home on 12 April after flying over 800 combat missions and expending over 659,000 lbs of ordnance. In May, its three A-1Js and nine A-1Hs were left behind at Cubi Point and VA-95 returned to NAS Lemoore, CA, where the Douglas A-4B Skyhawk was received in June 1965.

Above, VA-95 A-1J BuNo 142051 at NAS Quonset Pt., RI, in 1964 awaiting overhaul. (Jim Sullivan collection) Below, five VA-95 A-1Hs BuNos 137496 (NG/508), 135377 (NG/507), 139820 (NG/502), 135375 (NG/512), and 139810 (NG/501) and A-1J BuNo 142057 (NG/509) in flight from the USS Ranger (CVA-61) on 11 September 1964. (Jim Sullivan collection)

Attack Squadron Ninety-Six (VA-96) was established on 30 June 1956 at NAS Miramar, CA, and was equipped with AD-6 Skyraiders. Commanded by CDR Milton K. Dennis, VA-96 was part of Air Task Group Three (ATG-3) and was assigned to the USS Kearsarge (CVA-33). In August 1956, the squadron was transferred to NAS Moffett Field, CA.

On 4 June 1957, during the squadron's work-ups for its WestPac deployment, CDR Dennis was lost in a midair while operating off the Kearsage at night. The following day, CDR Stanley Sloan assumed command of the unit.

During the Kearsarge deployment from 9 August 1957 to 2 April 1958, the squadron operated off Taiwan in a show of force following the build-up of communist forces along the mainland opposite Taiwan. VA-96 also took part in "Operation Mantlerock" during the deployment. On return to CONUS with eight AD-6s, six AD-7s and one AD-5, the squadron was disestablished on 10 April 1958.

Above, VA-96 AD-7 BuNo 142075 taxis while folding wings after trapping. Note faded writing on the fuselage indicating aircraft was zapped by another carrier after the pilot landed on it. Aircraft has temporary wide orange wing and fuselage bands used during a war game exercise. (Paul Minert collection) Below, VA-96 AD-6 flown by ENS J.H. McKenzie being directed into the chalks by AO3 T.F. Neych, after competing in air-to-ground rocket firing at the Second Annual Naval Weapons Meet, El Centro, CA, on 2 April 1957. Note that the 20mm cannons have been removed. Tail stripes and wing tips were red/orange. (USN)

Above, VA-96 Sky-raider trapping on Kearsarge. (USN) At right, VA-96 AD-7 after trapping on CVA-33 in 1958. (USN) Below, VA-96 AD-6 BuNo 139750 at NAS Alameda, CA, in May 1958 after returning from its final deployment. Tail stripes, fin, and wing tips were red/orange. (William Swisher)

ATTACK SQUADRON ONE HUNDRED FOUR, VA-104 "HELL'S ARCHERS"

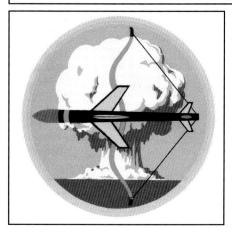

VA-104 was originally established as Fighter Squadron One Hundred Four (VF-104) on 1 May 1952 at NAAS Cecil Field, FL. Its initial aircraft was the FG-1D Corsair which was replaced by F4U-5s in December 1952. The squadron then moved to NAS Jacksonville, FL, in April 1953 and to NAS Cecil Field, FL, in December 1953 where

Above, VA-104 AD-6 BuNo 134585 launches from the Coral Sea in 1954. (Ginter collection) Below, VA-104 AD-6 BuNo 134584 at NAS Cecil Field on 9 June 1954. (NMNA) Bottom, VA-104 Skyraider (P/401) landing aboard the USS Leyte (CV-32) on 24 October 1955 during carrier qualifications. The squadron CO, CDR George E. Ford, is at the controls. (USN)

174

Above, VA-104 AD-6 BuNo 139723 in flight in 1959. (NMNA) Below, VA-104 AD-6 BuNo 139618 operates aboard the USS Ranger (CVA-61) during carrier qualifications in October 1958. (Hal Andrews collection)

that same month VF-104 was redesignated Attack Squadron One Hundred Four (VA-104). The "Hell's Archers" received their first Skyraiders (two AD-4Bs and sixteen AD-6s) in February 1954. That same month, CDR D.H. Johnson relieved LCDR Harold H. Brock as commanding officer. The squadron took their AD-6s aboard the USS Coral Sea (CVA-43) for a Mediterranean deployment from 7 July to 20 December 1954. A second Med deployment on the Coral Sea was conducted from 13 August 1956 to 11 February 1957 while under the command of CDR George E. Ford. CDR Jack N. Durio took over as CO in April 1957 and the squadron's "P" tail code was changed

to "AK" on 1 July 1957, prior to the squadron's last deployment. VA-104 was aboard the USS Forrestal (CVA-59) from 2 September 1958 to 12 March 1959 for its third Mediterranean cruise. On return to CONUS, the squadron was disestablished on 31 March 1959 at Jacksonville.

Above, VA-104 AD-6s BuNos 134551 (AK/411) and 134484 (AK/406) in flight from the USS Forrestal (CVA-59) on 10 October 1958. (USN) Below, VA-104 AD-6 BuNo 137546 refuels a VFP-62 Crusader. (USN)

ATTACK SQUADRON ONE HUNDRED FIVE, VA-105

Attack Squadron One Hundred Five (VA-105), commanded by CDR F.C. Auman, was established on 1 May 1952 at NAAS Cecil Field, FL. The squadron received fifteen AD-1s in May, their first two AD-4s in September and their first four AD-4NAs in October. The squadron was assigned to CVG-10 and the USS Tarawa (CVA-40) for a Med cruise from 7 January to 3 July 1953 with ten AD-4s and eight AD-4NAs.

Upon return to Cecil, LCDR R.S. Reeves assumed command in August 1953 and conducted training aboard CVA-45. In September and October, they boarded CVA-15 for carrier qualifications. Four AD-4Bs were added in October and the squadron operated from Cecil from November 1953 until July 1955 when they transferred to NAS Jacksonville, FL. In October 1954, CDR Samuel W.

At right, VA-105 AD-4 (P/507) landing aboard the USS Tarawa (CVA-40) in 1953. (USN) Below, VA-105 AD-4 BuNo 123799 at NAS Cecil Field in 1953. Wing tips and fin tip were green. (NMNA)

Forrer took command and VA-105 received its first two AD-6s the next month. By January 1955, the AD-4s were gone and sixteen AD-6s and one AD-5 were on-hand. The squadron was reassigned to ATG-201 and the USS Bennington (CVA-20). Carrier qualifications were conducted on CVA-20 in May and June and on 31 October 1955, a WestPac deployment began. It lasted through 16 April 1956 with the squadron returning to Cecil instead of Jacksonville. During the deployment, all fifteen pilots became Bennington centurions with 100 landings and the

Above, VA-105 AD-4 traps aboard CVA-15 off Guantanamo in October 1953. (NMNA) Below, VA-105 AD-4 (P/511) deck launches from CVA-15 in October 1953. Note location of wing codes and green wing tips and fin cap. (NMNA)

squadron garnished its second consecutive Atlantic Fleet Battle Efficiency "E" Award.

In June 1956, CDR Eugene F. Ternasky assumed command and preparations for another deployment

with ATG-201 began. This would not take place for another year and would be aboard the USS Essex (CVA-9) from 2 February through 17 November 1958. Early in the cruise, on 4 March, CDR Ternasky was killed during

Above, nine VA-105 AD-4s prepare for a deck launch from the USS Randolph (CVA-15) in October 1953 while operating off Guantanamo Bay, Cuba. (NMNA) Below, VA-105 AD-6 leads a squadron deck launch from the USS Bennington (CVA-20) as part of ATG-201 in 1955. Note "ATG" tail code. (Blair Stewart via Tailhook)

a night ditching attempt astern of the ship. CDR L.W. Cummins assumed command for the rest of the deployment. During July and August 1958, VA-105 flew close air support missions during the Marine landings in Beirut, Lebanon. Several aircraft were damaged by ground fire during their recon missions, but all returned. In September, the Chinese Communists began shelling the

Quemoy Islands and the Essex proceeded to the Western Pacific via the Med and the Suez Canal for duty in the Taiwan Straits.

In January 1959, the three CVG-17 AD squadrons (VA-42 on CVA-11, VA-104 on CVA-59 and VA-105 at Jacksonville) were reassigned to COMFAIRJAX. VA-105 was then disestablished on 1 February 1959 and

Above, VA-105 AD-6 BuNo 137596 from Essex refueling a VA-83 A4D-2 BuNo 142133. (USN) Below, two VA-105 AD-6s assigned to ATG-201 taxi on CVA-20 55-56. BuNo 137609 is in front. (Blair Stewart via Tailhook)

its eleven AD-6s and 120 personnel were absorbed by the AD RAG, VA-44 as part of CVG-4 at Jacksonville.

VA-114 was originally established on 10 October 1942 as Bombing Squadron Eleven (VB-11) at NAS San Diego, CA. VB-11 flew the Douglas SBD-3/-4/-5 Dauntless until November 1943 when it equipped with the Curtiss Helldiver. On 15 November 1946, VB-11 was redesignated Attack Squadron Eleven A (VA-11A) and then Attack Squadron One Hundred Fourteen (VA-114) on 15 July 1948.

VA-114 was assigned to CVA-45's CVG-11 and commanded by CDR Adolf L. Siegener in September 1948 when the squadron received its first two AD-1s. By the end of the following month, all the squadron's SB2C-5s were gone and the squadron began training operations up-and-down the coast of California. In January 1949, the squadron's fifteen AD-1s were replaced with fifteen AD-2s and weapons training was conducted at NAF El Centro, CA. On 30 June, LCDR J.E. Savage replaced CDR Siegener as CO and on July 4th

the USS Valley Forge was anchored off Santa Monica, CA, and open to the public. The squadron had an AD-2 on deck loaded with bombs and rockets and the public was allowed to sit in its open cockpit. On 1 August, LCDR E.T. Deacon relieved Savage as CO and two AD-3Qs were acquired later that month. The squadron was then scheduled for disestablishment and all Skyraiders

Above, fourteen VA-114 AD-2s in flight over San Diego in October 1949. (via Harry Gann) Below, VA-114 AD-2 BuNo 122365 aboard the USS Valley Forge (CV-45) on 4 July 1949 while anchored off Santa Monica, CA.. (William Swisher)

were transferred out in November. On 1 December 1949, VA-114 was disestablished.

ATTACK SQUADRON ONE - FIFTEEN, VA-115 "ARABS"

VA-115 was originally established as Torpedo Squadron Eleven (VT-11) on 10 October 1942 at NAS San Diego, CA, and equipped with TBF-1 Avengers. It was redesignated Attack Squadron Twelve A (VA-12A) on 15 November 1946 while still flying Avengers. VA-12A became Attack Squadron One Hundred Fifteen (VA-115) on 15 July 1948 at San Diego while assigned to CVG-11.

LCDR William H. House was the CO when the squadron received fifteen AD-2 Skyraiders to replace its TBMs in January 1949. On 3 August 1949, at NAAS El Centro, CA, during the Fleet's weapons meet, VA-115 became the Fleet's dive bombing champs. The winning team was composed of LCDR House, LT E.W. Gendron, ENS D.L. Miller, and ENS W.G. Sizemore. Also in August, two AD-3Qs were acquired and flown until November.

In December 1949, twelve AD-4s replaced all but three of the squadron's AD-2s and on 16 January 1950 LCDR Richard W. Fleck relieved CDR House. He took the squadron aboard the USS Philippine Sea (CV-47) on 24 July 1950 for the

VA-12A   VA-115

squadron's first Korean War deployment. They sailed with twelve AD-4s, two AD-4Qs and one AD-3Q.

On 5-6 August, VA-115 hit transformer stations at Chinju and Mangjin, power stations at Samehon and factories at Mokopo. From 12 to 15 September, the squadron struck targets in and around Inchon in preparation for the invasion. During and after the landings, they continued to strike targets around Inchon until 18 September. On 27 November 1950, AD-4 BuNo 123917 crashed into the ocean during a snowstorm. In December close air support was pro-

vided for the withdrawal from the Chosen Reservoir. Two aircraft were badly damaged on 9 December when AD-4 BuNo 123833 was damaged by a bomb blast and BuNo 123835 made a barrier crash aboard CV-47. AD-4 BuNo 123829 lost its engine on 16 December and crash landed at Yonpo. On 19 December, another

Below, VA-115 AD-4 BuNo 123841 in flight from the USS Philippine Sea (CV-47) on 10 June 1950. (National Archives)

Above, four VA-115 AD-4s in flight with their barn doors open on 6 October 1949. Fin tips were green. (National Archives) At right, VA-115's winning weapons meet team on 3 August 1949 with LCDR House reviewing the mission with LT Gendron, ENS Sizemore and ENS Miller. (USN) Bottom, bombed-up VA-115 AD-4s assigned to CV-47 inbound to Korea on 27 February 1951. BuNo 123851 has 50 mission marks on the fuselage side. Each aircraft was loaded with eight 250lb bombs and two napalm tanks. They also carry a centerline drop tank. (National Archives)

AD-4, BuNo 123915, was hit by bomb blast and ditched near Mayang Island

On 29 March 1951, while docked at Yokosuka, Japan, the squadron transferred to the Valley Forge for its return trip to CONUS. The squadron returned

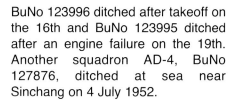

BuNo 123996 ditched after takeoff on the 16th and BuNo 123995 ditched after an engine failure on the 19th. Another squadron AD-4, BuNo 127876, ditched at sea near Sinchang on 4 July 1952.

After the war, VA-115 deployed twice aboard the USS Kearsarge (CVA-33) to the Western Pacific: from 1 July 1953 to 18 January 1954 with CDR J.D. Taylor commanding and from 7 October 1954 to 12 May 1955 with CDR C.L. Dillard commanding. During the second cruise, with sixteen AD-6s, the squadron flew air cover for the evacuation of more than 26,000 people from the Tachen Islands due to the Chinese shelling the islands.

The squadron transferred to the USS Essex (CVA-9) for its next WestPac cruise from 16 July 1956 to 26 January 1957 with CDR C.W. Smith Jr. commanding

Two WestPac deployments aboard the USS Shangri-La (CVA-38) followed. The first was from 8 March to 22 November 1958. VA-115 went to sea with CDR Leone Kirk Jr. in command who was relieved by LCDR R.L. Bothwell on 4 June 1958. CDR Bothwell remained CO for both cruises. During August and September 1958, VA-115 again aided the Nationalist Chinese after the Communist Chinese began shelling the Quemoy Islands. The second CVA-38 deployment was from 9 March to 3 October 1959. It was a world cruise beginning at San Francisco.

The Squadron's eighth AD deployment was aboard the USS Hancock (CVA-19) from 16 July 1960

At top, VA-115 AD-4 (V/503) is prepared for a mission from CV-47 on 19 October 1950. (Naval History) Below, VA-115 AD-6 BuNo 134635 on fire aboard the USS Kearsarge (CVA-33) on 4 August 1954. (National Archives) Bottom, bomb laden VA-115 AD-4 (V/516) deck launches from the USS Philippine Sea (CVA-47) in March 1952 while VF-113 and VF-114 F4U-4s wait their turn. (National Archives)

from Korea on 7 April and CDR Charles H. Carr relieved LCDR Fleck on 30 June. VA-115 deployed to Korea again aboard CV-47 from 31 December 1951 to 8 August 1952. AD-4 BuNo 127867 was shot down by AAA on 7 February 1952. An engine failure caused the ditching and loss of AD-4 BuNo 128919 on 9 March. In May, two AD-4Ls were lost.

At right, VA-115 officers 1954-55 left-to-right. On wing: ENS D.L. Hobson, ENS J.B. Marshall, CDR D.L. Dillard, and ENS R.A. Zick. Standing at center: LTJG R.R. Jarrell, LTJG J.F. Wanamaker, LTJG J.R. Emerson, LCDR J.R. Kincaid, LTJG J.E. Lott, LTJG C.A. Zimmerman, LTJG R.C. Maich, LTJG C.N. James, ENS J.R. Heath and LT W.J. Kwitkowski. Squatting: ENS W.A. Lott, LTJG R.F. Reynolds, LTJG L.S. Kollmorgen and LT D.L. Griewisch. Front row: LT F.F. Palmer, LTJG L.H. Gibson, LTJG P.F. Wattay and ENS F.M. Abbott. (USN) Below right, May 1955 deck crash of V/513 on CVA-33. (USN) Bottom, VA-115 AD-6 BuNo 135224 (V/502) taxis forward after trapping on Kearsarge in 1955. (USN)

to 18 March 1961. CDR G.W. Gaiennie was in command until 27 February 1961 when CDR C.H. Bowen took over.

Four USS Kitty Hawk (CVA-63) deployments were made by VA-115. The first was from 11 August to 1 November 1961 which included the carriers transfer to the Pacific via Cape Horn and included operations in the Caribbean and South Atlantic. The second CVA-63 cruise was from 13 September 1962 to 2 April 1963 with CDR George A. Parker assuming command on 3 January 1963. The third Kitty Hawk cruise took VA-115 to Indo-China. It was from 17 October 1963 to 20 July 1964 with CDR Merrill C. Pinkepank relieving CDR Parker on 21 January 1964. VA-115 flew CAP and SAR missions over Laos in May and June during the Laotian crises. During the squadron's last

Above, VA-115 AD-6 BuNo 139634 at NAS Miramar, CA, in 1956 while assigned to the USS Essex (CVA-9). (Warren Bodie) Below, VA-115 A-1H BuNo 135300 assigned to the USS Shangri-La (CVA-38) in 1957. Green tail, wing, and engine cowl trim. (Paul Minert collection) Bottom, VA-115 AD-6 BuNo 139705 taxiing in 1957. Aircraft was assigned to the CO, CDR L.F. Kirk Jr. (Jim Sullivan)

VA-115 "ARABS"

Above, VA-115 A-1H BuNo 139731 at NAS Moffett Field, CA, on 23 May 1962. (William T. Larkins) At right, VA-115 AD-6 BuNo 135251 with a Mk. 7 nuclear shape mounted on the centerline. (USN) Below, VA-115 AD-7 BuNo 142014 at MCAAS Yuma, AZ, on 3 December 1959. (William Swisher)

based paints. This resulted in the nickname of "Green Lizards" being used during the cruise. Four aircraft were lost, the first being A-1J BuNo 142038 on 1 February of 1966. LTJG B.S. Eakin bailed out his burning aircraft and was rescued after being hit over Laos. A-1J BuNo 142071 was lost at sea when its catapult bridle failed on 11 March 1966. The pilot was rescued. On 17 April 1966, LTJG William L. Tromp was lost in A-1H BuNo135398 on a night coastal armed recon flight. He had evaded a SAM near the Song Gia Hoi River and was returning to the ship when he had an emergency and crashed off shore. He was captured and died in captivity. The last loss was A-1J BuNo 142051, on 19 May 1966. An engine failure on takeoff put the aircraft in the sea. The pilot was quickly recovered.

The squadron's last deployment with the A-1 was from 5 January to 22 July 1967 aboard the USS Hancock (CVA-19) while assigned to CVW-5

Above, VA-115 A-1H BuNo 135373 being hoisted aboard ship. Fin tip was green. (Paul Minert collection) Below, VA-115 AD-7 BuNo 142012 taxis on the USS Shangri-La (CVA-38) in February 1960 during carrier qualifications. (USN)

CVA-63 deployment from 19 October 1965 to 13 June 1966, VA-115 took part in the Vietnam War. During this cruise, a number of the squadron's A-1H/Js were painted in experimental two tone green scheme with water-

Above, VA-115 AD-6 BuNo 134531 aboard the USS Kitty Hawk (CVA-63) at Yokosuka, Japan, on 16 October 1962. (Toyokazu Matsuzaki) At right, VA-115 officers aboard CVA-63 in 1963. (USN) Below, camouflaged VA-115 A-1H at Douglas in 1966 was assigned to the USS Kitty Hawk (CVA-63) in 1965-66. (Harry Gann)

and commanded by CDR H.G. Bailey. VA-115's mission was coastal patrol, gunfire spotting and SAR support. On 14 February, LT R.C. Marvin, veteran of 111 combat missions

VA-115 "ARABS"

Above, camouflaged VA-115 A-1J BuNo 135263 landing aboard the USS Kitty Hawk in the Tonkin Gulf in 1966. (Lionel Paul) At left, VA-115 A-1H BuNo 139778 in 1966 in its well worn temporary camouflage it wore while assigned to the USS Kitty Hawk (CVA-63). (Ginter collection) Below, VA-115 A-1H BuNo 135263 runs up prior to a mission in 1966. (USN)

Above, VA-115 A-1Hs BuNos 137612 (NF/504) and 135291 (NF/501) from CVA-19 enroute to targets in Vietnam in 1967. Fin tips were green. (via Tailhook) Below, BuNos 137601 (NF/506) and 134526 (NF/502). (via Tailhook)

aboard CVA-63, was lost. He had taken off on a RESCAP mission in A-1H BuNo 139805 and suffered engine failure. Both pilot and plane disappeared while attempting to return to the ship. Disaster struck the squadron on 17 March when four aircraft went down in two different incidents. LCDR A.H. Henderson in BuNo 135297 and LT R.B. Moore in BuNo 139768 were attacking a barge when they were hit by AAA. Both aircraft were ditched and both pilots were rescued. Then,

during a night SAR mission, the CO, CDR Bailey in BuNo 134625 and LTJG Gene W. Goeden in BuNo 135225, crashed off shore after a mid-air. Goeden died after cutting off Bailey's tail with his prop, but Bailey was later rescued. He had severe bruising of the head and a broken arm with exposed bone and was suffering from exposure.

In August 1967, the squadron stood down but was not disestab-

lished and was administratively assigned to VA-125. This is the only known instance that a Navy squadron was carried on the active books but as an inactive unit. It became active once again on 1 January 1970 as an A-6 intruder squadron.

While flying Skyraiders, VA-115 received a Presidential Unit Citation and Battle Efficiency E for 1960, 1962 and 1963.

SECOND ATTACK SQUADRON ONE-TWENTY-FIVE, VA-125 "ROUGH RAIDERS"

The second VA-125 was established as Attack Squadron Twenty-Six (VA-26) on 30 June 1956 at NAS Miramar, CA. It was originally equipped with F9F Cougars and carried the "Y"/"NC" tail codes of ATG-3. VA-26 was redesignated Attack Squadron One Hundred Twenty-Five (VA-125) on 11 April 1958 at NAS Moffett Field, CA. With the redesigna-

tion came a new mission. VA-125 became the Pacific Fleet A4D replacement squadron. To aide in this mission, an AD-5 was acquired in October 1960 to complement the squadron's sixteen A4D-1s and sixteen A4D-2Ns. A second AD-5 was subsequently acquired and the two AD-5/A-1Es were utilized into 1969 when the aircraft were retired at NAS

Above, VA-125 A-1E BuNo 132435 at NAS Lemoore, CA, in August 1965. Trim on Rough Raiders cargo tank was green. Second cockpit canopy was tinted blue. (Wiliam Swisher) Below, VA-125 A-1E BuNo 132466 in February 1969. (Ginter collection)

Lemoore, CA.

Attack Squadron One Hundred Twenty-Six (VA-126) was established in San Diego on 6 April 1956 with CDR G.L. Cassel commanding. It was originally equipped with Chance Vought F7U-3/3M Cutlasses. In April 1957, F9F-8/8Bs began replacing the F7Us and FJ-4Bs were received in 1958. The squadron became the FJ-4/4B RAG in 1958 and in April received twelve AD-6s, seven AD-7s and one AD-5 from the disestablished VA-54 and from VA-125 who discontinued the AD's usage in April. VA-126 Skyraiders wore an "NJ" tail code and sported 600 series aircraft numbers. On 30 April, there were twelve AD-6s, seven AD-7s, one AD-5, twelve FJ-4Bs, six F9F-8s and ten F9F-8Bs. In September, all the Cougars were gone and the first A4D-2s were acquired. By December, the Skyhawks were gone, too, and there

ATTACK SQUADRON VA-126

were only three AD-6s and the AD-5 left besides twenty-eight FJ-4/4Bs. The last of the Skyraiders were transferred out in January 1959. In 1960, VA-126 became the Pacific Fleet instrument training squadron with

F9F-8Ts and on 15 October 1965 was redesignated Fighter Squadron One Hundred Twenty-Six (VF-126).

ATTACK SQUADRON ONE - THIRTY - FIVE, VA-135 "THUNDERBIRDS"

Attack Squadron One Hundred Thirty-Five (VA-135) was established on 21 August 1961 at NAS Jacksonville, FL, with CDR Barclay W. Smith in command. VA-135 was formed as part of CVG-13 in response to the Berlin Crisis. The squadron had the distinction of receiving the Air Group's first aircraft, AD-6 BuNo 137497.

On 12 January 1962, three VA-135 Skyraiders became the eighth, ninth and tenth aircraft to trap aboard the USS Enterprise (CVAN-65). The squadron's only deployment was the shakedown cruise aboard the USS Constellation (CVA-64) to the Caribbean from March through May 1962. VA-135 transferred to NAS

Cecil Field, FL, in August 1962 where they were disestablished on 1 October 1962.

Above, VA-135 AD-6 BuNo 139797 traps aboard the USS Enterprise (CVAN-65) on 12 January 1962. (USN)

Above, VA-135 AD-6 BuNo 137497 (CVG-13's first air-craft) trapping aboard the USS Constellation (CVA-64) in February 1962. Fin tip was green. (USN) Below, VA-135 AD-6s on CVA-64 in March 1962. (USN)

Above, VA-135 AD-6 BuNo 139619 taxis aboard CVAN-65 on 17 January 1962. (USN) Below, VA-135 Skyraider AE/508 on the deck edge elevator of the USS Enterprise (CVAN-65). (via Tailhook)

FIGHTER SQUADRON ONE - FORTY - FOUR, VF-144 "BITTER BIRDS" ATTACK SQUADRON FIFTY - TWO, VA-52 "KNIGHTRIDERS"

VA-52 was originally established as Fighter Squadron Eight Hundred Eighty-Four (VF-884) at NAS Olathe, KS, on 1 November 1949. In response to the Korean conflict, VF-884 was called to active duty on 28 July 1950 and transferred to NAS San Diego, CA, where they equipped with F4U Corsairs. Two war cruises were made in 1951 and 1952-53 before the squadron reported to NAS Miramar, CA, in March 1953. In April 1953, the F9F-5 Panther was received followed by F9F-8/8B Cougars in April 1956.

In December 1958, while at Miramar, the squadron transitioned to the Skyraider. Both AD-5s and AD-6s

were used to man the squadron. On 23 February 1959, the squadron's mission was changed from fighter to attack and VF-144 was redesignated Attack Squadron Fifty-Two (VA-52). As such, VA-52 was assigned to CVG-5 for its next seven deployments. During 1959-60, weapons training was conducted at MCAAS Yuma, AZ, and NAS China Lake, CA, and carrier qualifications were aboard the USS Bennington (CVS-20) and the Ticonderoga. The first two West Pac cruises were aboard the USS Ticonderoga (CVA-14), the first from 5 March to 10 October 1960 and the second from 10 May 1961 to 15 January 1962. In March 1959, AD-7s were added and a mixed bag of AD-6/A-1Hs and AD-7/A-1Js were used during all the squadron's future deployments.

The squadron's third deployment was aboard the USS Lexington (CVA-16) for its transfer from the Pacific to the Atlantic Fleet from 21 July to 11 September 1962. At Cape Horn, six Skyraiders were launched for carrier qualifications, marking the first time any carrier-based aircraft had operated in this area known for foul weather and high seas. VA-52 then returned to the West Coast and Moffett Field,

CA, for its third deployment aboard the Ticonderoga from 3 January to 16 July 1963.

During the fourth cruise aboard CVA-14 from 13 April to 0 December 1964, VA-52 began its involvement with the conflict in Vietnam while under the command of George H. Edmondson. The Tico was part of

Below, VA-52 AD-6 BuNo 139622 at MCAAS Yuma, AZ, on 3 December 1959 for the Naval Air Weapons Meet. (William Swisher)

Above, VA-52 CAG bird, AD-5 BuNo 133858 at the Naval Air Weapons Meet in 1959. Note FAGU insignia on the aft fuselage. (Harry Gann via Jim Sullivan) At right, VA-52 A-1H BuNo 137537 taxis aboard CVA-14 on 4 August 1960. (USN) Below, VA-52 A-1J BuNo 142012 was assigned to the USS Ticonderoga (CVA-14) in 1963. Fin tip and drop tank markings were blue. (Paul Minert collection)

Yankee Team with three other attack carriers: the Bon Homme Richard, Constellation, and Kitty Hawk. Yankee Team was providing reconnaissance and SAR coverage over

Above, VA-52 AD-6 BuNo 135405. (Paul Minert collection) Below, VA-52 A-1H BuNo 137590 during gear retraction tests. Fin tip was blue. (Ginter collection) Bottom, VA-52 A-1H BuNo 139645 assigned to the USS Ticonderoga. Fuselage stripe and fin cap were dark blue. (Paul Minert collection)

South Vietnam and Laos and on 2 August, CVA-14 was the only attack carrier in the Gulf of Tonkin when the USS Maddox (DD-731) was attacked by three North Vietnamese P-4 PT boats. The Maddox was assigned to Desoto SIGINT patrol in international waters 28-miles off the coast when attacked. Three warning shots were fired, but when torpedos were launched, fire from the Maddox hit at least one PT boat and all torpedos were evaded. The Maddox, however, was hit by machine gun fire and called in air support from CVW-5. It arrived in the form of four F-8Es which sank one PT boat. During the night of 4-5 August, the Maddox and the USS Turner Joy (DD-951) were on patrol and identified radar contacts as probable torpedo boats and called for assistance. Two VA-52 Skyraiders flown by CDR Edmondson and LT Barton responded but were unable to locate the enemy patrol craft. The President authorized Operation Pierce Arrow, which called for limited air strikes against torpedo boats and support facilities in North Vietnam. The strikes were launched on the afternoon of 5 August with CVW-5 attacking the petroleum storage facilities at Vinh with a mixed bag of twenty-six A-1 Skyraiders, A-4 Skyhawks and F-8 Crusaders. After strike reconnaissance estimated that 90% of the

oil tanks were destroyed. VA-52 contributed four aircraft to the attack which were flown by CDR McAdams, LCDR Brumbach, LTJG Moore, and LTJG Carter. From 6 to 29 October, the squadron was responsible for combat rescue patrols in support of Yankee Team missions. For the deployment, VA-52 received the Navy Unit Commendation and the Armed Forces Expeditionary Medal. In total, the squadron was on the line 61-days before returning to NAS Alameda, CA. Upon return, the XO, CDR Lee T. McAdams, assumed command.

Above, VA-52 pilots aboard CVA-14 in 1960. (USN) Below, VA-52 A-1H BuNo 135332 (NM/301) landing at NAS Atsugi, Japan, on 1 November 1966. Aircraft was assigned to the USS Ticonderoga (CVA-14). (Toyokazu Matsuzaki)

VA-52 returned to Vietnam aboard the Ticonderoga from 28 September 1965 to 13 May 1966. Five Skyraiders were lost during the deployment, two were operational losses and three were combat losses. The operational losses were A-1H BuNo 137590 on 16 November 1965

Above, bombed-up VA-52 A-1H BuNo 135336 prepares to launch from CVA-14 for a mission over Vietnam in 1966-67. (USN) Below, VA-52 A-1H BuNo 134577 aboard the USS Ticonderoga. (Ginter collection) Bottom, VA-52 A-1H BuNo 137559 taxis forward while folding its wings in 1966-67 during operations while assigned to CVW-19 aboard CVA-14. Tail and fuselage trim were blue. (USN)

and A-1H 137621 on 1 December 1965. Both pilots were recovered safely. On 3 January 1966, LT J. W. Donahue was shot down by AAA over South Vietnam in A-1J BuNo 142081. Donahue was recovered safely, but the pilot of A-1H BuNo 139692 lost on 14 April 1966 was killed in action when he was hit with a SAM while attacking a wooden bridge over North Vietnam. He was CDR J. C. Mape, the squadron's CO since 10 December 1965. Mape was replaced with CDR Robert R. Worchesek on 19 April 1966. The last aircraft lost was A-1J BuNo 142032, which was shot down by AAA over North Vietnam. LT A. D. Wilson was recovered safely. During the cruise, LTJG Harvey M. Browne was awarded the Silver Star on 7 February 1966 for a rescue mission over Vietnam. Total days on the line for this cruise was 112.

CDR Worchesek took the squadron to Vietnam again aboard Tico from 15 October 1966 to 29 May 1967, this time as part of CVW-19. This would be the squadron's last deployment on the Tico and its last Skyraider cruise. During the 126-days VA-52 spent on the line, the squadron participated in Steel Tiger missions, Sea Dragon operations, coastal reconnaissance, and combat rescue air patrol missions. Steel Tiger missions were concentrated ground attacks in Southern Laos and Sea Dragon operations involved spotting for naval gunfire against coastal radar, gun battery sites and cargo craft. During the deployment, CDR John F. Wanamaker received the Silver Star on 9 March 1967 and only two Skyraiders were lost. On 27 November 1966, LTJG W. H. Natter was recovered from North Vietnam after being shot down by AAA in A-1H BuNo 135341. Then, on 18 January 1967, a VA-52 pilot was killed in an operational accident in A-1H BuNo 139748.

On 30 June 1967, one month after returning to Alameda, CDR Lester W. Berglund, Jr. took command. The next day the squadron was ordered to NAS Whidbey Island, WA, where they transitioned to the Grumman A-6A Intruder.

ATTACK SQUADRON ONE - FIFTY - TWO, VA-152 "FRIENDLIES"

VA-152 was originally established at NAS Denver, CO, as Reserve Fighter Squadron Seven Hundred Thirteen (VF-713) in 1950. Due to the Korean War, the squadron was called to active duty on 1 February 1951 and equipped with Chance Vought Corsairs. In April 1951, the squadron transferred to NAS Alameda, CA. Two F4U-4 combat cruises were made aboard CV-36 and CVA-37 before the squadron transitioned to the McDonnell F2H-3 Banshee in October 1953. While aboard CVA-37, the squadron was redesignated Fighter Squadron One Hundred Fifty-Two (VF-152) on 4 February 1953. Four Banshee WestPac deployments followed aboard CVA-10, CVA-18, CVA-12, and CVA-20 before being replaced by Skyraiders. On 1 August 1958, VF-152 was redesignated Attack Squadron One Hundred Fifty-Two (VA-152).

In January 1959 at NAS Moffett Field, CA, VA-152 received its first Skyraider, an AD-5 which they operated alongside their seven F2H-3s.

By the end of February, strength was three F2H-3s, one AD-5, and eleven AD-6s. By the end of June, all F2H-3s were gone and sixteen AD-6s and one AD-5 were on-hand.

LCDR V.E. Sanderson took command from CDR Royce A. Singleton on 16 February 1959 and took the squadron on its first Skyraider deployment as part of CVG-15. On 1 August 1959, the USS Hancock (CVA-19) sailed for a WestPac deployment, returning on 18 January 1960. After return to CONUS, LCDR R.M. Sullivan became the acting CO on 8 February at Moffett Field. CDR John

A. Davenport relieved Sullivan on 4 March 1960.

Davenport took the squadron aboard the USS Coral Sea (CVA-43) from 19 September 1960 to 27 May 1961 for its second WestPac deployment. In January 1961, they operated in the South China Sea in response to

---

Below, VA-152 AD-6 BuNo 137502 from the USS Coral Sea (CVA-43) in 1961. (USN)

Above, Coral Sea-based VA-152 A-1J BuNo 142014 landing at NAS Atsugi, Japan, on 19 September 1963. (Toyokazu Matsuzaki) Below, VA-152 AD-6 BuNo 134605 over the USS Coral Sea (CVA-43) on 22 March 1961. (USN via Barry Miller) Bottom, VA-152 AD-6 BuNo 139776 trapping aboard CVA-43 on 4 March 1961. (USN)

the crisis in Laos.

After returning to Moffett, CDR John R. Bicknell assumed command on 21 June 1961. VA-152 deployed again aboard CVA-43 from 12 December 1961 to 17 July 1962. During the cruise, on 11-12 July the squadron participated in the first Navy flight operations since WWII in the Bearing Sea.

A third CVA-43 cruise was conducted from 3 April to 25 November 1963 with CDR R.B. Bergner in command of VA-152.

CDR H.F. Gernert assumed command on 3 March 1964. On 22 April, the squadron deployed to Bien Hoa Air Base, South Vietnam, to train South Vietnamese pilots in the A-1.

Det Zulu was maintained at Alameda until 1 August when VA-152 returned to Alameda and a small Det Zulu was maintained at Bien Hoa.

On 19 March 1965, CDR Albert E. Knutson took over and deployed the unit as part of CVW-16 aboard the USS Oriskany (CVA-34) from 5 April to 16 December 1965 on the squadron's first war cruise to Vietnam. VA-152 arrived on station on 8 May and began operations over South Vietnam. Then, on 18 June their operations shifted to North Vietnam and Laos which included a four-plane Det to Udon AB, Thailand,

for SAR support. Two operational losses due to engine failures occurred on 30 June (A-1H BuNo 139708) and on 21 July (A-1H BuNo 139636). Another Spad, A-1J BuNo 142012, and its pilot LTJG Lawrence S. Mailhes were lost near Tiger Island to unknown causes during a rescue mission on 10 August. On 26 August, A-1H BuNo 139720 was lost due to AAA during a night armed recon mission. LTJG Edward A. Davis became a POW after bailing out and was repatriated in January 1973. Another plane and pilot were lost on 29 August when LT Edd D. Taylor was killed by AAA fire while flying SAR

Above, VA-152 AD-6 taxiing forward on the USS Coral Sea (CVA-43) on 1 March 1961. (National Archives) Bottom, VA-152 A-1H BuNo 139728 off Vietnam with Zuni rocket pods and ace of spades insignia on 31 October 1966. (USN)

coverage for an F-105 rescue. On 7 November LT Gordon Wileen was forced to make a gear-up landing at Da Nang in A-1H BuNo 134563 after being hit by AAA on an F-105 SAR mission. LCDR Paul Merchant's A-1H

BuNo 137566 was hit by AAA on the night of 9 November and forced to ditch off the coast. Another gear-up landing was made at Da Nang by LT Clarke after a SAR mission. LCDR Jesse J. Taylor died during a SAR mission for a VA-163 A-4E pilot when A-1H BuNo 135244 was hit by AAA on 17 November. He was able to exit the target area but crashed 15 miles southwest of Haiphong. Taylor's wingman, LCDR Schade, was also hit by AAA and his A-1H was written-off after a hard landing on CVA-34. VA-152 started the deployment with twelve Skyraiders and lost thirteen during the cruise with replacements flown in from Cubi Point as needed.

VA-152 conducted two more war cruises aboard CVA-34 from 26 May to 16 November 1966 and from 16 June 1967 to 31 January 1968. CDR Gordon H. Smith, the XO in 1965, was in command for the squadron's second war cruise. Operations from Dixie Station began on 30 June. Nine days later, VA-152 was operating from Yankee Station. On 7 August, LT Charles W. Fryer was lost while attempting to ditch A-1H BuNo 139701 after being struck by small arms fire while attacking a train. During an armed recon mission on 18 August, LCDR Eric Schade fired a couple of Zunis into a suspicious wooded area which resulted in secondary explosions. His wingman, LT

Above, VA-152 A-1J BuNo 142501 at NAS Moffett Field, CA, on 16 October 1964 while assigned to the USS Oriskany (CVA-34). (William T. Larkins) Bottom, VA-152 A-1H BuNo 137502 at NAS North Island, CA, while assigned to the USS Oriskany (CVA-34) on 28 April 1966. (William Swisher)

A.J. Garvey, joined in with Zunis and 20mm, then six more VA-152 Skyraiders arrived to unload on the trucks, too. The de-nuded clearing was dubbed "Eric's Truck Park". Two A-1Hs were lost on 25 August, one on launch. BuNo 135236 flown by LTJG James A. Beene ditched due to the

failure of a catapult holdback hook. He was rescued only to perish on 5 October when he crashed at sea while transiting a thunderstorm in A-1H BuNo 137610. The second aircraft BuNo 135231 was lost by the CO, CDR Smith. He was hit by AAA and bailed out of his burning aircraft over the Tonkin Gulf. He dropped out of the A-1H while inverted and struck the tail, but was recovered safely by helo. On 8 October, the carnage continued with LT John A. Feldhaus being hit in A-1H BuNo 137629. He subsequently crashed and died. His wingman, LTJG Quenzel, was also hit but was able to return to the ship. Another pilot, ENS Darwin J. Thomas in A-1H BuNo 139731, was lost at night during a rocket firing run. A hangar deck fire on Oriskany claimed the life of the XO, CDR John Nussbaumer and AZAN David A. Liste, on 26 October 1966.

CDR D.M. Wilson was in command for the squadron's last Skyraider cruise in 1967-68 aboard CVA-34. The squadron began operations on 14 July and on the 15th LTJG Robin B. Cassell was killed on an armed recon mission along the coast of Vietnam in A-1H BuNo 135288. SAR support, fleet gunfire spotting and coastal patrols were the order-of-the-day during the cruise. The squadron was credited with 132 watercraft including PT boats during the deployment.

After returning to CONUS, VA-152 immediately started to convert to the Douglas Skyhawk. The last two A-1Hs were retired in February 1968.

Above, Zuni-equipped VA-152 AD-6 BuNo 135370 in flight off Vietnam while operating from the USS Oriskany (CVA-34) in 1967. (USN) Bottom, VA-152 A-1H BuNo 134518 with 105 mission marks on the fuselage side. Tail chevron was yellow outlined in black. (Paul Minert collection)

VA-154 was originally established as Bombing Squadron One Hundred Fifty-Three (VB-153) on 26 March 1945 at NAAS San Manteo with Curtiss Helldivers. By August 1946, the squadron was operating from NAS Alameda, CA, and was redesignated Attack Squadron Fifteen A (VA-15A) on 15 November 1946. They deployed once with their SB2C-5s from 31 March to 8 October 1947 before receiving their Douglas Skyraiders. On 15 July 1948, VA-15A was redesignated Attack Squadron One Hundred Fifty-Four (VA-154) and twelve AD-2s were received in August 1948. The twelve AD-2s were flown along with five SB2C-5s through January 1949. The squadron operated for short periods at sea aboard the USS Antietam (CV-36) in October and December 1948 and aboard the USS Boxer (CV-21) in January and February 1949 with sixteen AD-2s for carrier training. LCDR Charles N. Conatser was in command throughout the squadron's employment of the Skyraider. All aircraft were transferred out in November and the squadron was disestablished on 1 December 1949.

Above, VA-154 AD-2 on CV-36 in 1948. Prop hub and fin cap were yellow. (USN) At top right, VA-154 (A/412) and VA-155 (A/501) AD-2s over San Francisco Bay on 2 December 1948. (USN) Bottom, VA-154 AD-2 aboard Antietam CV-36 in 1948. Fin tip was yellow. (USN) Above right, VA-154 and VA-155 AD-2s in flight over San Francisco, CA, in 1948. (William T. Larkins)

THE FIRST ATTACK SQUADRON ONE - HUNDRED - FIFTY - FIVE, VA-155

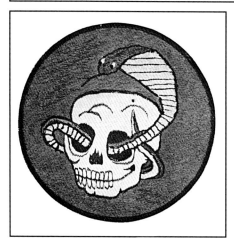

The first VA-155 was originally established as Torpedo Squadron One Hundred Fifty-Three (VT-153) at NAAF Lewiston on 26 March 1945. With TBM Avengers, the unit transferred to NAAS Oceana, FL, on 1 June 1945. A second duty-station transfer occurred on 2 July to NAS Norfolk, VA, followed by a final transfer to NAS Alameda, CA, on 8 August. On 15 November 1946, VT-153 was redesignated Attack Squadron Sixteen A (VA-16A). LCDR Gerald R. Stablein was in command when on 15 July 1948 VA-16A was redesignated Attack Squadron One Hundred Fifty-Five (VA-155). The squadron received twelve AD-2s in August and boarded the USS Antietam (CV-36) for training in September. They returned to

Below, VA-155 AD-2 BuNo 122225 at NAS Alameda, CA, on 21 September 1948 assigned to the CO of VA-155, LCDR G.R. Stableim. Fin cap was green. (William T. Larkins)

VA-155 AD-2 BuNo 122231 drops its left wing, clipping the deck, which added to the engine torque momentum and the aircraft goes over the side, rotates inverted, and impacts the water that way on 31 August 1948 while operating from CV-45. Note the debris in the water forward of the main gear. The pilot failed to get out. (National Archives via Jim Sullivan)

Antietam in October and in December for further carrier operations. In January and February 1949, VA-155 operated off-and-on from the USS Boxer (CV-21) with their sixteen AD-2s. LCDR Don L. Ely replaced Stablein in January 1949 and would dicostablish the squadron on 30 November 1949.

ATTACK SQUADRON ONE - SIXTY - FIVE, VA-165 "BOOMERS"

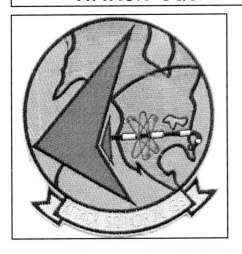

Attack Squadron One Hundred Sixty-Five (VA-165) was established

at NAS Jacksonville, FL, on 1 September 1960 under the command of CDR Carl H. Yeagle. The squadron's first aircraft were two AD-6 Skyraiders which were received in October. The AD-6s were acquired slowly with five on-hand in November, six in December, ten in January 1961 and twelve in June. On 25 August, CDR John E. Ford assumed command and on 7 September, VA-165 was transferred to NAS Moffett Field, CA.

On 7 June to 17 December 1962, the squadron deployed for the first time. CDR R. Houck took VA-165's eleven AD-6s and one AD-7 aboard the USS Oriskany (CVA-34) as part of Air Group Sixteen for a WestPac cruise. A second WestPac cruise aboard Oriskany was conducted from 1 August 1963 to 10 March 1964. During the cruise, CDR Houck was

---

Below, VA-165 CAG bird, AD-6 BuNo 139701, on 22 May 1962. (William T. Larkins) Bottom, VA-165 A-1H BuNo 135374 at NAS Moffett Field, CA, in 1964 while assigned to the USS Oriskany (CVA-34). (William Swisher)

relieved by CDR L.L. Andrews, Jr. on 2 September 1963. In November 1963, the unit operated off the coast of South Vietnam during the tension created by the overthrow of President Diem. After returning to CONUS, CDR R.E. Chamberlain, Jr took command and VA-165 was reassigned to Air Wing Fifteen on 22 June 1964.

The Squadron's third deployment was their first war cruise to Vietnam. It was conducted aboard the USS Coral Sea (CVA-43) from 7 December 1964 to 1 November 1965 (the longest carrier deployment of the Vietnam War). During the cruise, CDR A.K. Knoizen relieved CDR Chamberlain and even though they

participated heavily in the Rolling Thunder campaign only two aircraft were lost. The first, A-1H BuNo 139772, was lost on 13 August 1965 after being hit by ground fire during an armed recon mission. LT R.J. Hyland nursed the stricken aircraft over the shore where he bailed out and was rescued by a USAF HU-16B. On 4 September, during another armed recon mission, LTJG Edward B. Shaw was killed by AAA in A-1H BuNo 139693.

After the squadron returned to Alameda, it was reassigned to CVW-10 in preparation for another Vietnam deployment. The Champagne Air Wing (CVW-10) was the first all attack Wing of the war. It was equipped with

Above, VA-165 A-1H BuNo 139765. Note unusual circle style Attack Squadron 165 application. (NMNA) Bottom, VA-165 A-1H BuNo 137532 launches from CVA-34 on 12 November 1962. (USN)

VA-15 and VA-95 flying A-4 Skyhawks and VA-165 and VA-176 flying Skyraiders. They were loaded aboard the USS Intrepid (CVS-11) on 4 April 1966 at NAS Norfolk, VA, which had been classified a limited Attack Carrier for the deployment. With CDR Harry D. Parode in command, they sailed to the Tonkin Gulf via the Mediterranean and the Suez Canal. The squadron's first strike took

Above, Coral Sea-based VA-165 A-1H 134562 in 1964-65 with green trim. (Paul Minert collection) Below, VA-165 A-1H BuNo 134504 was assigned to the USS Intrepid (CVS-11). Fin tip was yellow. (Paul Minert collection) Bottom, VA-165 A-1H BuNo 135338 assigned to the squadron's CO, CDR Harry Pardoe, for the 1966 war cruise aboard the USS Intrepid (CVS-11). Bomb below Pardoe's name shows 138 missions. Note Tonkin Gulf Yacht Club insignia on the aft fuselage. (William Swisher)

place on 15 May 1966. Two aircraft were lost during the deployment; the first, A-1H BuNo 137534, belonged to XO CDR William S. Jett. He was hit by AAA during an armed recon mission and bailed out over the gulf where he was rescued by chopper. The

Above, VA-165 A-1H BuNo 135305 on the deck of the USS Intrepid (CVS-11) in May 1966. Aircraft belonged to the XO, CDR Bill Jet. (Jim Sullivan collection) Below, VA-165 A-1H BuNo 135272 AK/210 about to launch from the Intrepid in 1966. Fin tip was yellow, and fuselage boomerang was green. (USN)

event was repeated by LTJG T.J. Dwyer on 13 September in A-1H BuNo 134534. Both pilots had been operating near Vinh.

After returning to CONUS on 21 November 1966, CDR Jett took command on 22 December. VA-165 was transferred to NAS Whidbey Island, WA, on 1 January 1967 where they transitioned to Grumman A-6A Intruders.

Above, VA-165 A-1J BuNo 142059 "Puff the Magic Dragon" from CVS-11 at Da Nang in 1966 with 55 mission marks on the fuselage. Fin tip was yellow and fuselage chevron was green. (NMNA)

ATTACK SQUADRON ONE - SEVENTY - FOUR, VA-174 "BATTERING RAMS"

VA-174 was orignally established as Bombing Squadron Eighty-Two (VB-82) at NAS Wildwood on 1 April 1944. Their first aircraft were SB2C Helldivers which they operated until March 1948 when they were replaced with AM-1 Maulers. VB-82 was redesignated Attack Squadron Seventeen A (VA-17A) on 15 November 1946 at NAS Quonset Point, RI. Then, on 11 August 1948, VA-17A was redesig-

At right, VA-174 AD-3 piloted by LTJG Farley exposes its belly detail with open dive brakes on 28 June 1949 near Cecil Field, FL. (USN)

nated Attack Squadron One Hundred Seventy-Four (VA-174). In February 1949, the squadron transferred to NAAS Cecil Field, FL, where they received sixteen AD-3s in April 1949 while under the command of LCDR Robert E. Farkas. On 10 June, LCDR

William R. Pittman relieved LCDR Farkas and in July the last AM-1 Maulers were retired. In October, carrier qualifications were conducted on the USS Midway (CVB-41) as part of CVG-17. The squadron finished December with sixteen AD-3 and was

Below, VA-174 AD-3s on 28 June 1949. R/410 was piloted by LTJG Farley, R/416 by ENS T.E. Daum, and R/412 by ENS G.W. Lockwood. (USN)

disestablished on 25 January 1950.

Above, VA-174 AD-3s in flight on 28 June 1949. Plane Captain names are written on the engine cowls. (USN) Below, two views of VA-174 AD-3 BuNo 122791 while taxiing at NAS Floyd Bennett Field, NY, in 1948. (NMNA)

VA-175 was originally established as Torpedo Squadron Eighty-Two (VT-82) at NAS Quonset Point, RI, on 1 April 1944. The squadron equipped with TBM Avengers and flew five different versions before receiving Skyraiders in February 1949. VT-82 was redesignated Attack Squadron Eighteen A (VA-18A) on 15 November 1946. Then, VA-18A became Attack Squadron One Hundred Seventy-Five (VA-175) on 11 August 1948. VA-175 was commanded by LCDR John E. Kennedy when the first AD-3 was received in February 1949. By the end of March, fourteen AD-3s were on-hand.

The squadron was assigned to CVG-17 and prepared for a deployment aboard the USS Coral Sea (CVB-43) to the Med from 9 September 1950 to 1 February 1951. In April 1950, the AD-3s were

Above, two VA-175 AD-3s aboard CVB-42 in 1950. (Tailhook) Below, VA-175 AD-3 aboard the USS F.D. Roosevelt (CVB-42) in 1949. Fin tip trim was white. (USN)

replaced with AD-4s of which sixteen were in use during the cruise. In May and in June, carrier qualifications were conducted aboard CVB-43.

The squadron's next deployment

with AD-4s and AD-4Ls was aboard the USS Franklin D. Roosevelt (CVB-42) from 3 September 1951 to 4 February 1952 again to the Med. On return to NAS Jacksonville, FL, LCDR Ross A. Knight relieved LCDR Malcolm E. Wolfe in May 1952 and began preparing for another CVB-42 deployment.

The third deployment was to the North Atlantic and the Mediterranean aboard CVB-42 from 26 August to 19 December 1952 with two AD-4Ls and fourteen AD-4s.

At right, VA-175 AD-4s run-up on the aft flight deck of CVB-43 in November 1950. (USN) Below, VA-175 AD-4 BuNo 123879 noses over aboard the USS Coral Sea (CVB-43) during a barrier crash on 14 October 1950. Pilot was LTJG Fritze. (USN)

For the squadron's fourth AD cruise, they sailed aboard the USS Wasp (CVA-18) for an around-the-world cruise from 16 September 1953 to 1 May 1954. After returning to CONUS, VA-175 began replacing its AD-4s with AD-6s in August 1954.

CDR W.L. Nyburg was in command for the squadron's fifth Skyraider deployment. VA-175 found itself aboard CVA-43 again on its way to the Med from 5 April to 29 September 1955. During the Suez crisis, an emergency deployment off the coast of Spain took place in November and December 1956.

For the squadron's last deployment while under the command of CDR E.H. Potter, Jr., they boarded the FDR for the third and final time. The Mediterranean cruise was from 12 July 1957 to 5 March 1958.

VA-175 was disestablished on 15 March 1958 at NAS Jacksonville, FL.

At top left, VA-175 AD-4B BuNo 132348 takes the barrier on CVA-18 on 14 October 1953. (USN) At left, VA-175 AD-6 BuNo 139805 on CVA-42 off Guantanamo on 26 September 1956. (NMNA) Bottom, VA-175 AD-6 in flight from the USS Coral Sea (CVA-43) in 1955. Prop hub, fin tip, rudder edge, and horizontal tail tips were white. (Tailhook)

ATTACK SQUADRON ONE - SEVENTY - SIX, VA-176 "THUNDERBOLTS"

Attack Squadron One Hundred Seventy-Six (VA-176) was established on 1 June 1955 at NAS Cecil Field, FL. VA-176 was assigned to Air Task Group 202 (ATG-202) and was commanded by CDR James M. O'Brien, Jr. The unit had nine AD-6s on-hand at the end of July 1955 and would fly the AD-6/A-1H until February 1969 when replaced by Grumman A-6A Intruders. In February 1956, VA-176 was transferred to nearby NAS Jacksonville, FL.

During the training for their first deployment, the squadron also oper-ated an AD-4N alongside their fourteen to sixteen AD-6s. Carrier training operations aboard the USS Randolph (CVA-15) were conducted in March through June 1956 during its shake-down cruise. The squadron's first deployment was aboard CVA-15 from 14 July 1956 to 19 February 1957. During this period, because of the tensions of the Suez War, operations were concentrated off the coast of Egypt from October through December 1956.

After the squadron returned to Jacksonville, CDR L.S. Cummins took command on 10 April 1957. Also in April an AD-5 was acquired to aid in training. February 1958 found the unit operating out of Guantanamo

Above, two bombed-up VA-176 AD-6s with ATG-202's "X" tail code prepare to take off from the USS Randolph (CVA-15) on 25 August 1956. (National Archives) Below, VA-176 AD-6 BuNo 139614 operates aboard the USS Randolph (CVA-15) in 1956 as part of ATG-202. Rudder, fin and fuselage lightning bolt were red. (USN)

Bay, Cuba. In March 1958, VA-176 was assigned to CVG-17 and CDR D.C. Standley took over as CO. 18 April 1958 found VA-176 participating in the two-week Naval Air Weapons Meet at NAAS El Centro, CA. The Thunderbolts were reassigned to CVG-10 in July 1958. They returned

Above and below, VA-176 competitors at the 1958 Naval Air Weapons Meet at NAAS El Centro, CA, on 18 April. The aircraft carried an "AL" tail code at the time and had orange wing and fin tips and orange prop hubs. (Harry Gann)

to Gitmo for training from January to April 1959. Carrier qualifications were accomplished in May through July aboard the USS Essex (CVA-9) in preparation for a second Mediterranean deployment. CVA-9 and VA-176 deployed from 7 August 1959 to 26 February 1960 with CDR U.W. Patrick in command.

A quick turnaround had VA-176 aboard the USS Shangri-La (CVA-38) for a North Atlantic deployment from 6 September to 20 October 1960. With CDR Robert J. Stegg in command, CVA-38 and VA-176 were sent to South America following a request from Guatemala and Nicaragua to help prevent infiltration of communist guerrillas from Cuba. VA-176 was on station aboard Shangri-La from 14-25 November, then cross-decked to the USS Wasp (CVA-18) while underway and remained on station until 8 December 1960. While shore-based in January 1961, the squadron had use of two AD-5s and one AD-5Q. On 2 February 1961, VA-176 was back aboard the "Shang" with twelve AD-6s and one AD-5 while headed for the Med. They returned to Jacksonville on 15 May 1961 but were ordered to the Dominican Republic from 2-19 June 1961 in response to the assassination of GEN Rafael Trujillo. CDR B.B. Forbes took command on 11 September 1961.

VA-176 was back aboard CVA-38 again on a Med deployment from 7 February to 28 August 1962 with CDR H.P. Maulden in command.

From April to June 1963, VA-176 and CVA-38 were called to operate in the Caribbean during a crisis in Haiti and the Dominican Republic. CDR R. Brooke took over on 27 September 1963 and took the squadron back aboard CVA-38 for another Med cruise from 1 October 1963 to 23 May 1964.

After returning to CONUS, CDR George D. Edwards, Jr. took command on 9 October 1964 and took the squadron to the Med from 15 February to 20 September 1965, again aboard CVA-38.

In 1966, while commanded by CDR Robert J. Martin, CVW-10 was reorganized into the first all-attack Air Wing of the Vietnam War. It was comprised of VA-15 and VA-95 flying A-4 Skyhawks and VA-165 and VA-176 flying Skyraiders. Nicknamed the Champagne Air Wing, (CVW-10) was loaded aboard the USS Intrepid (CVS-11) on 4 April 1966 at NAS Norfolk, VA. They sailed to the Tonkin Gulf via the Mediterranean and the Suez Canal. The squadron's first strike took place on 15 May 1966.

Only one pilot, LT Charles A. Knochel, was lost to combat in

VA-176

Above, two VA-176 AD-6s BuNos 139806 (AK/410) and 139745 (AK/403) conduct live fire rocket training in March 1959. (USN) At right, A-1H CAG bird from VA-176, BuNo 134536. Trim was orange. (Paul Minert collection) Below, VA-176 A-1H BuNo 137624 armed with two 250 lb bombs, three rocket pods, and a four-shot Zuni pod on the right wing. (Paul Minert collection)

Vietnam. On 22 September 1966, he was hit in A-1H BuNo 135239 as he went feet wet after an armed recon mission. He bailed out over the water but drowned after hitting the water. On 27 September, CDR A.R. Ashworth relieved CDR Martin during their third line period.

The squadron's most momentous day was on 9 October 1966 when

LTJG William T. Patton downed a MiG-17. Patton was part of a RESCAP flight led by LCDR Leo Cook. They were covering a massive three-carrier Alpha strike near Hanoi and after an F-4 went down 20 miles southwest of Hanoi, Cook's flight went feet dry and raced in. Cook's wingman was LTJG Jim Wiley and Patton's was LTJG Russell. They encountered heavy flak on the run-in

Above, VA-176 A-1H BuNo 139634 assigned to Intrepid at Baltimore, MD, in 1964. Note large chipped squadron insignia on fuselage side. (Dave Lucabaugh) Below, VA-176 A-1H BuNo 135326 being serviced for an air strike against the Viet Cong on 20 May 1966 aboard CVS-11. An ordnancman is loading a silver Mk 77 mod 1 napalm bomb, two crewmen are fueling the bird and one crewman is topping off the oil tank aft of the engine. (USN)

Above, VA-176 A-1H BuNo 137496 assigned to LCDR W. Zimmerman being fueled on the USS Intrepid (CVS-11) in September 1966. (USN) Below, VA-176 A-1H BuNo 134622 taxis after trapping on CVS-11 (Paul Minert collection)

VA-176

Above, VA-176 A-1H 134472 at NAS Los Alamitos, CA, in late 1966 after its Intrepid war cruise. (William Swisher) At left, flak shredded tail of BuNo 135239 aboard CVS-11. (USN) At right, LTJG William T. Patton poses on the wing of Skyraider adorned with a MiG kill marking after the incident. (USN) Below, MiG killer, LTJG William T. Patton stands with Douglas officials in front of A-1H BuNo 135326 at NAS Los Alamitos, CA, in late 1966. Douglas public relations executive Harry Gann is at left. (Harry Gann)

for the first few minutes, then suddenly all AAA fire stopped. This was followed by Cook informing the flight that there were four MiGs in the area. The Skyraiders were on their own with no fighter CAP to keep the MiGs at bay. A short while later, Wiley broadcast "I've got three MiGs taking turns on me! Please get some fighter cover in here!" To which Patton replied "Right "Pud", wer'e on our way with two "Spads" and a helo!" There were four MiGs; one went after Cook and the other three after Wiley. Cook was able to evade, while Wiley was down low trying to shake the MiGs. He banked hard to the left around a mountain while one of the MiGs broke right around the mountain and then cut left, which put him in front of Wiley who opened up with his 20mm. He shredded the MiG's wing tip which then departed the area venting a vapor trail. Wiley would be credited with a probable kill.

It was then that Patton and Russell arrived and took Wiley's other two pursuers on a head-on pass. Russell fired at the nearest MiG and was credited with a probable kill. Patton then spotted another MiG low over the trees and broke into a dive,

rolling out at the MiG's four o'clock and firing. The MiG attempted a reverse turn into Patton which failed to put him on Patton's tail; instead, it put Patton on the MiGs "six". The MiG then headed skyward at about 75° but had lost much of his speed and was unable to escape Patton as he fired all four guns until they would fire no more. The 20mm shells shredded the MiG's tail and peppered the engine. Patton then fired three Zunis in an attempt to finish the job. The Zunis missed and the MiG flipped and entered a cloud heading for the ground. Patton gave chase and fired one more Zuni which also missed right before Patton entered the cloud. When Patton exited the cloud at 500 ft he saw the pilot eject. For the action, Patton was awarded the Silver Star and Cook, Wiley, and Russell received DFCs. After the war, it was learned that only Patton's MiG went down; the other three made it back to base.

During the deployment, VA-176 flew 1,606 combat sorties and dropped 2,566 tons of ordnance and was credited with 15 trucks, 17 rail cars, 3 AAA flights, 84 barges and 14 bridges. After returning to CONUS,

VA-176 was reassigned to CVW-3 and the USS Saratoga (CVA-60).

The squadron's final Skyraider deployment was to the Mediterranean aboard CVA-60 from 2 May to 6 December 1967. The USS Liberty (AGTR-5) was mistakenly attacked by Israelis on 8 June 1967 and four squadron aircraft were launched to assist the ship. Israel apologized for the attack and the aircraft were recalled shortly after launch.

During the cruise, on 1 August 1967, CDR J.T. French relieved CDR Ashworth. Back at Jacksonville on 25 April 1968, CDR Charles L. Cook took command and only three A-1Hs were on-hand at the end of the month. In May, when all its Skyraiders were gone, the squadron transferred to NAS Oceana, VA, and the transition to the A-6 Intruder began.

Below, VA-176 A-1H 134472 assigned to XO CDR A.R. Ashworth (later CO) made a wheels-up landing and took the barrier aboard the Intrepid on 19 July 1966. (USN via Jim Sullivan)

VA-176

Above, VA-176 A-1H BuNo 139702 at Maxwell AFB in April 1968 after the squadron's final Spad deployment on CVA-60. (Barry Miller) At left, VA-176 A-1H BuNo 139810 in 1967 after assignment to CVW-3 and the USS Saratoga (CVA-60). (Tailhook) Below, VA-176 A-1H BuNo 137524 aboard the USS Saratoga (CVA-60) in 1967. Fin and wing tips were orange. (USN)

FIGHTER SQUADRON ONE - NINE - FOUR, VF-194 "YELLOW DEVILS"
ATTACK SQUADRON ONE - NINE - SIX, VA-196 "MAIN BATTERY"

VA-196 was originally established as Fighter Squadron One Hundred Fifty-Three (VF-153) on 15 July 1948 at NAS Alameda, CA. On 15 February 1950, VF-153 was redesignated Fighter Squadron One Hundred Ninety-Four (VF-194). Equipped with F8F Bearcats, they deployed once in 1950 aboard the USS Boxer (CV-21) before receiving F4U-4s in August 1950.

In December 1950, while under the command of LCDR Robert S. Schreiber, the squadron began receiving AD-3 Skyraiders. Receipt of the ADs was slow with only two AD-3s and five AD-1s on-hand to augment the squadron's twenty-five F4U-4s in February 1951. In June, the Corsairs were finally gone and VF-194 was operating eighteen AD-1s.

The squadron's first AD deployment with five AD-3s and eleven AD-2s took place aboard the USS Valley Forge (CV-45) from 15 October 1951 to 3 July 1952 as part of ATG-1. On

11 December 1951, the squadron's first combat missions were flown against rail lines and bridges. Interdiction of the rail and road network in North East Korea would remain the squadron's mission throughout the cruise. On 18 December, the XO, LCDR Ben Pugh, was forced to ditch near Wonsan due to damage received during a railcutting mission. He successfully exited the aircraft but succumbed to expo-

Below, VF-194 AD-3s with BuNo 122763 (B/401) in the, foreground pose with the Air Groups Bearcats. (Paul Minert collection)

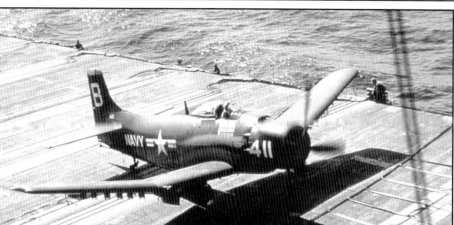

Above, VF-194 AD-3 BuNo 122753 taxis at Alameda in 1951. (Paul Minert collection) At left, VA-194 Skyraider taxis forward while folding wings on CV-45 after a mission over Korea. Note yellow mission marks on the fuselage side. (USN) Below, VF-194 AD-3 deck launching in 1951. Fin tip was yellow. (Via Bob Lawson)

sure. On 1 January 1952, LCDR Schrieber led a highly successful strike against barracks in the Wonsan

area. Another ditching took place on 8 January 1952. LTJG A.A. Peterson was hit by small arms fire during a rail strike north of Hungnam and was safely rescued after 45 minutes in the water. Two more aircraft were lost in February. On the 3rd, LTJG N.J. Johnson was hit in the accessory section by a 37mm round and bailed out over enemy lines. He was rescued safely 30 minutes later by a helo while squadronmates kept the North Koreans at bay with their 20mm guns. Two days later, LT John C. Workman was also hit by flak in the accessory section but was able to ditch off Wonsan before being picked up unharmed 10 minutes later by a destroyer. On the 8th, ENS Marvin Broomhead was hit while attacking a bridge and was forced to crash land nearby. A Navy helo attempted to res-

At top right, VF-194 AD-4NA BuNo 125750 off the Korean coast in June 1953 with its bomb load supplemented with a yellow survival canister. Aircraft has a yellow lightning bolt below the canopy and fin cap. (NMNA) At right, VF-194 AD-4N BuNo 125761 in flight from CVA-21 in 1953. (Paul Minert collection) Below, VF-194 AD-4NL BuNo 125762 taxis aboard CVA-21 on 3 June 1953. (USN)

229

cue the badly injured pilot but also crashed. Then a failed rescue attempt by a USAF helo was attempted and all three men were captured. During the deployment, CAG Crabill led ATG-1 in a VF-194 Skyraider. One such raid on 18 February had the ADs loaded with a 2,000lb bomb and 250lb bombs or two 1,000lb bombs and 250 pounders to strike seven high-priority targets at Pungsan. The ATG-1 Skyraiders and Corsairs accounted for destruction of 45 buildings and 38 damaged buildings. On 22 March, ENS K.A. Schechter was blinded by an exploding shell hitting his canopy. LTJG J.H. Thayer joined up and through his series of verbal commands guided the blinded pilot to a successful wheels-up landing at K-50. Another VF-194 pilot, LTJG J.P. Cooper, was also hit in the canopy and had his helmet holed, but was able to land safely at K-10. CAG Crabill was also forced to ditch on the 18th and was rescued by the Ptarmigan (AMS-376). LT John Workman was hit again on the 19th and died in the low-level bailout from his burning aircraft. On the 20th, LCDR D.E. Brubaker received a grazed left temple when a small arms round entered the cockpit. After combat operations on 10 June 1952, five squadron ADs were flown to K-3 for delivery to the Marines and three more were given to Princeton's Air Group. On the eleventh, the remainder of the ADs were transferred to the FASRON 11 pool at Atsugi.

After return to CONUS, the squadron was assigned a small number of F6F-5s temporarily before LCDR Arthur N. Melhuse assumed command on 19 August 1952. Soon, eight AD-1s replaced the Hellcats and permanent AD-4NAs began arriving. By November there were sixteen AD-4NAs on-hand.

In March, with eighteen AD-4NAs and two AD-4Qs, a second war cruise

VF-194 AD-4Q BuNo 124055 (B/419) went off the deck of CVA-21 on 5 July 1953. LTJG Joe Kagi climbs out before the aircraft sank. (National Archives via Jim sullivan)

was conducted, this time aboard the USS Boxer (CVA-21) from 30 March to 28 November 1953. Seven Skyraiders were lost during the cruise. While off Molokai on 13 April, BuNo 127000 suffered a prop failure and was successfully ditched by the pilot. Korean gunfire claimed BuNo 125756 on 17 May which ditched off-shore near Tanchon. AD-4NA BuNo 127005 was lost on 20 May after a bad takeoff from Boxer. Enemy AAA claimed AD-4NA BuNo 126919 near Hungnam on 23 May. AD-4NA BuNo 125754 was ditched after takeoff from CVA-21 on 15 June. On 5 July, LTJG Joe Kagi put AD-4Q BuNo 124055 in the water off the deck of CVA-21 and was safely rescued. Another AD-4NA, BuNo 126950, ditched after takeoff from CVA-21 due to a power failure on 11 July. The Boxer operated off Korea from 10 May to 21 June and after the Armistice on 27 July conducted training operations in the Sea of Japan until 11 November 1953.

In December 1953, LCDR B.R. Trexler assumed command and AD-6s were received to replace the squadron's mixed bag of AD-4s. Carrier qualifications aboard the USS Wasp (CVA-18) were conducted in August 1954 and the squadron deployed to the Western Pacific aboard CVA-18 from 1 September 1954 to 9 April 1955. In February 1955, the squadron provided air cover during the evacuation of the

Above, VA-196 AD-4 BuNo 137534 aboard CVA-16 on 22 May 1956. Although trim should have been orange, it was orange-yellow because of VF-194/VA-196's first nickname the "Yellow Devils". (SDAM) Below, two VF-194 Skyraiders over CVA-18 in 1954. (USN)

231

Tachen Islands.

On 4 May 1955, VF-194 was redesignated Attack Squadron One Hundred Ninety-Six (VA-196). While at Alameda, an AD-5 was rovolved to aid in training. In June, CDR R.B. Giblin took over and began preparing the squadron for its next AD-6 deployment. In May 1956, VA-196 boarded the USS Lexington (CVA-16) for a WestPac cruise from 28 May to 20 December 1956.

The squadron's fifth AD cruise and last ATG-1 deployment was aboard the USS Ticonderoga (CVA-14) from 4 October 1958 to 17 February 1959 with LCDR Dicky Wieland commanding. During the cruise, VA-196 received the CNO

Safety Award and visited Okinawa, Subic Bay, Manila, Hong Kong, and Yokosuka.

Five AD/A-1 deployments aboard the USS Bon Homme Richard as part of CVG/CVW-19 followed. The three pre-Vietnam WestPac cruises were from 21 November 1959 to 14 May 1960, 26 April to 13 December 1961, and from 12 July 1962 to 11 February 1963.

The fourth BHR deployment began in peacetime on 28 January 1964. On 2 August 1964, the Tonkin Gulf Incident occurred. As a result, the BHR was ordered to Vietnam from the Indian Ocean to take part in Yankee Team retaliatory strikes. While commanded by CDR J.R.

Above, VA-196 AD-6 BuNo BuNo 134564 traps aboard the USS LEXING-TON (CVA-16) as part of ATG-1 on 18 July 1956. (National Archives) Below, VA-196 AD-6 on CVA-16 in 1957. Prop hub, wing tips and fin tip were yellow outlined in black. (USN) At top right, VA-196 AD-6 BuNo 137495 assigned to ATG-1 aboard the USS Ticonderoga (CVA-14) in 1958. (William T. Larkins) At right, VA-196 AD-6 BuNo 135294 assigned to ATG-1 aboard the USS Ticonderoga (CVA-14) in 1958. Fin tip was orange boardered in black. The star and the number 1 on the fuselage side was bordered in orange. (William T. Larkins) Bottom right, VA-196 AD-6s over San Francisco Bay on 26 September 1958 while assigned to ATG-1. Fore-to-aft; BuNo 134634 (NA/402), 135319 (NA-404), 134561 (NA/401), and 135278 (NA/403). (Hal Andrews via Jim Sullivan)

Above, VA-196 AD-6 BuNo 134522 runs-up on 16 May 1959 prior to taxiing with bombs and rockets on the wing stations and a Boar nuclear shape on the centerline. (Swisher collection) Below, VA-196 AD-6 BuNo 139652 landing on CVA-31 on 15 May 1960. (Ginter collection) Bottom, VA-196 AD-6 BuNo 139765 from the USS Bon Homme Richard (CVA-31) landing at NAS Atsugi, Japan, on 8 August 1961. (Toyokazu Matsuzaki)

Driscoll, these strikes were flown from 31 August to 8 October. The squadron returned home on 21 November 1964 and CDR Joseph Gallagher became CO on 4 January 1965.

Three A-1Js and eight A-1Hs were deployed on the second combat cruise from 21 April 1965 to 13 January 1966. During the deployment, four aircraft and three pilots were lost. The first three aircraft were lost in September 1965. The first to go down was A-1J BuNo 142057 which was destroyed when its pilot, LCDR James Kearns, was killed by his own bomb dropped on 14 September. On the 24th, the CO, CDR Gallagher, bailed out in-country after being hit by small arms fire in A-1H BuNo 135274. He was later rescued by a Marine helicopter. The third September loss was LCDR Carl J. Woods and A-1H BuNo 134482 on the 28th. He was hit by AAA while attacking a railway line, made it to the coast, where he bailed out of his burning aircraft and drowned. The fourth aircraft, A-1H BuNo 139755,

Above, VA-196 AD-6 BuNo 137545 with orange fin tip and drop tank markings outlined in black. (Stephen Miller) Bottom, four VA-196 A-1Hs BuNos 139776 (NM/514), 139638 (NM/512), 139702 (NM/513), and 134482 (NM/507) over NAS Lemoore, CA. (USN)

and third pilot LCDR Gerald Roberts, were lost on 2 December 1965. It is believed he flew into the ground when hit on his third attack run on a bridge 35 miles north of Dong Hoi.

Above, VA-196 A-1H BuNo 137537 at NAS Lemoore with colorful orange trimmed drop tank on 19 July 1963. (Doug Olson) At left, VA-196 Skyraider launches from CVA-31 for a mission over Vietnam in 1965. Fin tip was orange. (USN) Bottom, VA-196 A-1H NM/604 landing aboard CVA-31 on 5 March 1964. (USN)

After return to CONUS, CDR James A. Donovan took command and the squadron transitioned to the A-6A Intruder prior to its next Vietnam deployment. The last Skyraiders were transferred out in February 1966.

1956

1958

1959

Attack Squadron Two Hundred Fifteen (VA-215) was established on 22 June 1955 at NAS Moffett Field, CA. The first CO was CDR E.E. Kerr and the first Skyraider, an AD-5, was received in July. Nine AD-6s were added in August and eight more in September.

CDR Kerr took VA-215 aboard the USS Bon Homme Richard (CVA-31) for its first deployment on 16 August 1956. The "Barn Owls" were assigned to CVG-21 for its first WestPac cruise. In October, during a three carrier strike exercise, the squadron ran afoul of Typhoon Jean. Bad weather caused cancellation on the first day and even though it improved little during the night, the squadron made a pre-dawn launch the following morning. Join up was extremely difficult due to 65 knot winds, a pitch black night and a 250 foot ceiling. The unit's mission was to make a simulated special weapons attack using the loft-bombing method. By the time they reached the target (a cluster of small rocks sticking out of the sea), conditions had worsened and one section even flew through Jean's eye. After the simulated release, the unit which was operating under radio silence headed back to the ship, a 1,000 mile flight round trip. Thankfully, the ceiling had improved for the landing cycle and all sixteen aircraft were recovered on the badly pitching and rolling deck. Only one airframe was damaged beyond the ship's ability to repair it. The Squadron returned to Moffett on 28 February 1957 and LCDR W.A. Skon took command in March 1957.

CVG-21's and therefore VA-215's "G" tail code was changed to "NP" on 1 July 1957, prior to its second WestPac deployment from 14 July to 19 December 1958 aboard the USS Lexington (CVA-16). CDR O.N. Ford was in command for this cruise in-and-around Taiwan. On return to

---

Below, VA-215 AD-6s off Southern California while assigned to the USS Bon Homme Richard (CVA-31) in 1956. (USN)

CONUS, LCDR Peter Rippa took command in January 1959, only to be replaced temporarily by LCDR J.L. Morrison, Jr., on 25 March 1959 after Rippa was killed on a training flight in Yosemite National Park. CDR H. Broadbent, Jr., took over on 3 April 1959. CDR Broadbent took VA-215 back aboard the "Lady Lex" for another WestPac deployment from 26 April to 2 December 1959. Ports-of-call were: Hawaii, Yokosuka, Hong Kong, Iwakuni, Beppu, Subic Bay, Okinawa, and Sasebo.

Back at Moffett, CDR G.A. Buckowski relieved CDR Broadbent on 4 April 1960 and prepared VA-215 for a third AD cruise on CVA-16. They deployed once again to the Western Pacific from 29 October 1960 to 6 June 1961. Overseeing crises on Formosa and in Laos kept the squadron busy during the cruise.

In 1962, the first of four USS Hancock (CVA-19) deployments began on 2 February. Under the command of CDR L.A. Dewing, VA-215 operated off the coast of South Vietnam in April to cover the arrival of the first Marine advisor unit from the US. In May, they were off to Thailand due to tensions along the Thai/Laotian border. Then in June, they operated off of Quemoy and Matsu Islands due to tensions in that region. Also in June, CDR F.W. Brown, Jr., relieved CDR Dewing. The first Hancock deployment ended on 24 August 1962.

The second CVA-19 deployment was under the command of CDR R.J. Licko and was from 7 June to 16 December 1963. Once again, tensions between Taiwan and Communist China called for VA-215's presence in September.

A third CVA-19 deployment occurred from 21 October 1964 to 29 May 1965 with CDR Donald D. Brubaker in command. Continuing problems with North Vietnamese activities in Laos authorized operation Barrel Roll (an armed reconnaissance program) in Southern Laos. Strikes began on 14 December with VA-215 flying their first strikes on 28

Above, VA-215 AD-6 BuNo 137604 launching from CVA-31. (Craig Kaston collection) Below, VA-215 Skyraiders run-up prior to launch from CVA-16 in 1958. For the cruise, the squadron sported orange checkerboard tails. (USN) Bottom, VA-215 AD-6 BuNo 137596 on CVA-16's elevator in 1958. (USN)

December 1964. In February 1965, attacks on US barracks both at Pleiku and Qui Nhon prompted retaliatory strikes called Flaming Dart I & II, of which VA-215 took an active roll. On 1 March 1965, CDR Robert C. Hessom relieved CDR Brubaker and the squadron began Rolling Thunder operations. A-1H BuNo 139790 flown by LTJG C.E. Gudmunson was hit by AAA while attacking a radar site on 26 March. He crash-landed safely within 5-miles of Da Nang and was rescued. On the 31st, LTJG Gerald W. McKinley perished to AAA while attacking another radar site in A-1H BuNo 137584. On 2 April, LCDR James J. Evans died in BuNo 139721 along the Ho Chi Minh Trail when his plane went in. LTJG James P. Shea flew A-1H BuNo 139818 into the ground on 19 April while attacking a truck convoy, bringing VA-215's death toll to three. The unit's fifth air-craft loss occurred on 27 April when LTJG S.B. Wilke's A-1H BuNo 137545 was hit in the wing and caught fire. He bailed out 2-miles short of Nakhon Phanom AB, Thailand, and was safely recovered.

The last CVA-19 deployment was another war cruise to Vietnam. It was from 10 November 1965 to 1 August 1966. Initial operations were over the Mekong Delta and of flying support for Marine Operations Jackstay and

Above, VA-215 emergency landing on CVA-61 resulted in lots of graffiti. (USN) Below, VA-215 AD-6 BuNo 135366 over CVA-19 in 1963. (USN) Bottom, VA-215 A-1J BuNo 142081 runs-up with a nuclear shape under the centerline. (USN)

239

Deckhouse. A series of losses similar to their first war cruise plagued this one, too. The first loss was that of the CO, CDR Hessom, on 5 March in A-1H BuNo 137500. He had been hit by AAA south of Vinh during a Rolling Thunder strike and had coached the aircraft for 15-miles before it crashed and killed him. CDR Frederick L. Nelson assumed command for the remainder of the cruise. Then, during a Steel Tiger armed recon mission on 29 April, LCDR William P. Egan was shot down and killed in A-1H BuNo 139616 in Laos. On 9 May, LCDR C.W. Sommers II ditched off-shore and was rescued during a small-boat patrol in A-1H BuNo 139616. This was followed on 14 May by the loss of A-1J BuNo 142050 during launch due to a control system failure. After return to Lemoore, CDR George A.

Carlton took command in January 1967.

The squadron's last Skyraider deployment with two A-1Js and eight A-1Hs was aboard the USS Bon Homme Richard (CVA-31) from 26 January to 25 August 1967. During the cruise, two aircraft and pilots were lost. The first loss was LT Paul C. Charvet in A-1H BuNo 137516, who crashed at sea near Hon Me Island on 21 March. The second loss was ENS Richard C. Graces who crashed during a rocket run in A-1H BuNo 135366 on 25 May. Upon return to CONUS, VA-215 was disestablished on 31 August 1967.

Above, VA-215 A-1H BuNo 139770 waiting to deck launch from the USS Hancock (CVA-19) on 24 March 1965. (USN via Barry Miller) Below, VA-215 A-1J BuNo 142063 deck launches from the USS Hancock. (USN)

ATTACK SQUADRON TWO - SIXTEEN, VA-216 "BLACK DIAMONDS"

Attack Squadron Two Hundred Sixteen (VA-216) was established on 30 March 1955 at NAS Moffett Field, CA, with CDR Frank Ault in command. In April, two AD-4s and two AD-4NAs were received and in May an AD-5 was added. Eight AD-4Bs and eight AD-4NAs deployed on the USS Yorktown (CVA-10) from 19 March to 13 September 1956 to the Western Pacific. After returning to Moffett, CDR Hope Strong, Jr., relieved CDR Ault on 30 September 1956.

In October 1956, all the AD-4Bs and AD-4NAs were transferred out and six AD-7s and one AD-5 were received. CDR Strong took thirteen AD-7s aboard the USS Hornet (CVA-12) from 6 January to 2 July 1958 for another WestPac deployment. Upon its return to CONUS, the squadron did an immediate turnaround due to the Quemoy and Matsu crisis arriving in the Formosa Straits in September aboard the USS Bennington (CVA-20). On 12 January 1959, they

returned to Moffett and CDR W.E. Payne, Jr., became CO on 13 February 1959 with ten A4D-2s replacing all the squadron's Skyraiders by month's end.

Below, VA-216 AD-4N with BuNo 125722 (G/613) in the foreground followed by AD-4B 132306 (G/608), AD-4N 126910 (G/610) and AD-4B 132248 (G/607) while assigned to CVA-10 on 7 July 1956. (National Archives)

Above, VA-215 AD-7 BuNo 142034 trapping aboard CVA-12 in 1958. (USN) Below, VA-216 AD-7 BuNo 142029 with 142030 behind it. Tail markings were orange. (Ginter collection) Bottom, VA-216 AD-7 BuNo 142029 at NAS Moffett Field, CA, while assigned to ATG-4 in June 1957. (Larry Smalley)

244

ATTACK SQUADRON SEVEN - ZERO - TWO, VA-702 "RUSTLERS"
ATTACK SQUADRON ONE - FORTY - FIVE, VA-145 "SWORDSMEN"

Naval Reserve Attack Squadron Seven Hundred Two (VA-702) was established on 1 December 1949 at NAS Dallas, TX, and equipped with TBM Avengers. The first skipper, LCDR S.C. Seagraves, was in command on 20 July 1950 when VA-702 was called to active duty because of the conflict in Korea. The following month they relocated to NAS San Diego and in August received eighteen AD-2 Skyraiders. An AD-4Q was added in September and a second one in January 1951.

VA-702 took eleven AD-2s and two AD-4Qs aboard the USS Boxer (CV-21) for their first Korean War deployment from 2 March to 24 October 1951 as part of CVG-101. First combat operations began on 27 March. On 11 May, sixteen VA-702 and sixteen VA-195 Skyraiders struck railway bridges along the west coast in response to an Air Force request. Each AD carried two 2,000 lb bombs and VA-702 dropped three spans of one bridge and one span of another. On return to ship, the pilot of AD-2, BuNo 122303, spun-in on approach to Boxer and was recovered with minor injuries. Also in May, a replacement aircraft, AD-4 BuNo 123818, and its pilot were lost on the 7th when they failed to pull out of a dive. On 21 June, AD-2 BuNo 122213 was hit by AAA and was lost.

Once back at San Diego, CDR

Bruce T. Simonds took command on 6 December 1951. The demand for ADs in Korea left the squadron with only four AD-2s and two interim AD-1s. By February 1952, sixteen AD-1s and three AD-2s were being used for training and in April VA-702 began receiving its replacement aircraft, one AD-4L and three AD-4s. By the end of July, all the AD-1s were gone and six AD-4Ls and ten AD-4s were in operation. The squadron's second Korean War cruise took place aboard the USS Kearsarge (CV/CVA-33) from 11 August 1952 to 17 March 1953 as part of CVG-101, once again with combat operations commencing on 14 December. Tragedy struck five days after sailing when the skipper, CDR Simonds, crashed off the bow in AD-4 BuNo 123965 immediately after launch. LCDR Harry C. McClaugherty assumed command for the remainder

of the cruise. During the cruise, VA-702 was redesignated Attack Squadron One Hundred Forty-Five on 4 February 1953. On the 8th, AD-4 BuNo 123871 was shot down by AAA. Also in February, ENS Bill Doggett's AD-4 was hit over "Artillery Valley" by a 37mm round. It penetrated a prop blade leaving a hole the size of a grapefruit and foot-long streamers of metal. Shrapnel from the hit left over 100 holes in the engine and right wing. With his extremely vibrating, bucking and smoking engine threatening to shake the plane

---

Below, VA-702 AD-2 BuNo 122343 near Japan while operating off the USS Boxer (CV-21) in September 1951. Crudely written "Tiger 17" and "Briny Marlin II" adorn the forward fuselage. (Naval History)

Above, VA-702 AD-2 "Shook II" over Korea with a 2,000 lb bomb on 25 April 1951 during the attack on the Changjin River Bridge. (USN) Above right, ENS Doggett holds his prop blade that was holed by a 37mm shell in February 1952. (USN) Bottom, VA-702 AD-2 122313 flown by LTJG R. Smith makes the 47,000th landing on CV-21. (USN) At right, VA-702 AD-4 deck launches from CV-33 in August 1952 loaded six with 5" HVARs, six 250 lb bombs and two 500 lb bombs for targets in Korea. (USN) At right bottom, VA-145 AD-4L BuNo 132308 crash landed on the USS Essex (CVA-9) on 5 October 1953. (USN)

apart, he still made it safely to an auxiliary field some 100 miles away. The plane was repaired by replacing the prop, engine, hydraulics and repairing the holes in the skin.

After returning to San Diego and transfer to NAS Miramar, CDR John A. Duncan took command. During training at El Centro, its 14 pilots bailed out of a R4D in mass to find out what it felt like. Carrier qualifications were conducted in October 1953 aboard the USS Essex (CVA-9) and in January 1954 aboard Kearsarge. On 3 February 1954, CVG-14 with VA-145 reported to the Atlantic Fleet for temporary assignment aboard the USS Randolph (CVA-15) for a Med cruise ending on 6 August 1954. While on cruise, VA-145 was award-

ed the Air Pac VF/VA Prop Safety Award. VA-145 took part in Operation Hellenic Sky I from 25 February to 2 March, Operation Shield I from 26 March to 4 April, and Italic Sky I from 30 April to 4 May 1954. The squadron made fourteen port calls before returning to Miramar

From Miramar, two weapons training periods were conducted at El Centro and carrier qualifications were conducted aboard CVA-19 and CVA-21. After which, the squadron's fourth

deployment and second aboard CVA-21 was conducted from 3 July 1955 to 3 February 1956. LCDR/CDR Gale L. Bergey deployed with eight AD-4Bs and eight AD-4s. In September 1955, VA-145 set a fleet-wide all-time flight hour record with 1,097.4 hours flown.

Upon return, CDR W.P. Blackwell took command in February and AD-6s began replacing the AD-4s. March 1956 finished with thirteen AD-6s, one AD-5, and two AD-4Bs on strength, and by June all the AD-4Bs

Above, VA-145 AD-4Bs at NAS Denver, CO, in 1955. (NMNA) Below, VA-145 AD-4 BuNo 123786 runs up for launch from CVA-21 on 20 April 1955. Note squadron insignia on the engine cowl. (Paul Minert collection)

were gone. During this period a squadron pilot was lost in training. Twelve AD-6s deployed aboard the USS Hornet (CVA-12) from 21 January to 25 July 1957. During the

cruise, the CO lead a section of VA-145 Skyraiders over Swatow, China, by mistake which resulted in LT Bill Steger's plane being holed by communist Chinese flak.

Above, VA-145 AD-6 BuNo 137567 from the USS Hornet (CVA-12) in flight over the Philippines in February 1957. (National Archives) Below, VA-145 AD-6 BuNo 134571 off the USS Hornet (CVA-12) in 1957. Trim was green. (Paul Ludwig) Bottom, VA-145 AD-6 BuNo 139701 at NAS Miramar, CA, on 10 August 1957 after its return from a deployment aboard the USS Hornet (CVA-12). During the deployment on 1 July 1957, the unit's "A" tail code was changed to "NK". (Doug Olson)

Above, VA-145 AD-6 BuNo 139633 at NAAS El Centro, CA, in April 1958. Tail trim was orange. (USN via Peter Mersky) At left, VA-145 pilots pose during the 1958 Naval Air Weapons Meet at El Centro on 18 April. (Harry Gann) Bottom, four VA-145 AD-6s begin their break into NAAS El Centro, CA, during the 14 April 1958 3rd Annual Fleet Air Weapons Meet. (USN)

After returning to CONUS in July 1957, CDR Charles S. Brooks assumed command and in preparation for its next cruise competed in the 1958 Naval Air Weapons Meet at El

Centro in April. CDR W.H. Alexander II took over on 15 September 1958 and on 3 January 1959, VA-145 boarded the USS Ranger (CVA-61) at NAS Norfolk, VA, for its around-the-Horn transfer from the Atlantic to the Pacific Fleet which ended on 27 July 1959. During 5 to 8 July, VA-145 operated off Taiwan during increased tensions in the area.

In August 1959, CDR Harvey S. Herrick assumed command and was relieved by CDR Warren H. Ireland on 11 May 1960 in time to take ten AD-6s aboard the newly modernized USS

Oriskany (CVA-34) on 14 May. CDR Herrick reassumed command on 17 August and the deployment lasted until 15 December 1960.

Back at Miramar, CDR E.B. Berger took the helm on 28 December 1960. He was relieved by LCDR B.L. Blackwelder on 12 September 1961 and CDR R.A. Norin on 12 October 1961. During this period, weapons training was conducted at Yuma as were carrier qualifications aboard CVS-12 and CVA-16 and an ordnance demonstration aboard CVA-61. On 9 November 1961,

Above, VA-145 A-1H BuNo 135406 on 3 December 1959 at NAAS El Centro, CA, while assigned to the USS Ranger. (William Swisher) Bottom, VA-145 A-1J BuNo 142074 aboard the USS Constellation (CVA-64) with two others at Yokosuka, Japan, on 26 April 1963. (Toyokazu Matsuzaki)

eleven VA-145 AD-6s boarded the USS Lexington (CVA-16) for a WestPac deployment ending on 12 May 1962. During the cruise, VA-145 was known as "Norin's Turtles" as

they visited ten ports-of-call.

After relocation to NAS Moffett Field, CA, in May 1962, two USS Constellation (CVA-64) deployments followed: from 21 February to 10 September 1963, and from 5 May 1964 to 1 February 1965. After the first cruise the squadron relocated to NAS Alameda, CA, across the bay. Prior to the first deployment, CDR Blackwelder assumed command on 11 October 1962 and was followed by CDR H.A. Hoy during the cruise on 31 July 1963. Hoy was still in command for the start of the second Deployment but was relieved at sea by CDR Melvin D. Blixt on 3 August 1964. VA-145 joined Yankee Team operations off Vietnam in June then Connie made port call in Hong Kong. They were recalled to Vietnam due to the Tonkin Gulf incident of 2 August when three PT boats attacked the USS Maddox. Two days later, the Maddox (DD-731) and the Turner Joy (DD-951) thought they had been attacked again and President Johnson authorized retaliatory strikes on the 5th. VA-52 launched from Tico and VA-145 launched two strikes from Connie at long range as the ship was still not on station. LTJG Richard C. Sather in the second strike attacked two Swatow PT boats near Lach Thuong when he was hit by AAA. He died in A-1H BuNo 139760 when he crashed two miles offshore and became the first Naval aviator lost to enemy action in what would

become the Vietnam War. A second VA-145 A-1H was also hit by AAA during the strike, but returned to ship.

A second Vietnam deployment was made aboard the USS Ranger (CVA-61) from 10 December 1965 to 25 August 1966 with CDR H.F. Griffith in command. VA-145 arrived on Dixie Station on 14 January 1966, flying missions in support of US and ARVN forces in South Vietnam.

The unit's strikes moved north on the 31st when CO Griffith led a foul weather strike of six Spads which destroyed two highway bridges and two staging areas. After coordinating the attacks, the unit received a SAR request for a downed VA-55 pilot near Dong Hoi. During the mission, Griffith was hit twice, rolled inverted, recovered and made it safely to an auxiliary field. For his actions he was awarded the Silver Star. On 1 February, LTJG Dieter Dengler was shot down by ground fire in A-1J BuNo 142031 during a SAR mission for a VA-115 pilot. He was captured and imprisoned in Laos, but escaped on 29 June and was eventually rescued by a USAF

HH-3E on 20 July. For his exploits he was awarded the Navy Cross.

LTJG Gary D. Hopps was killed in A-1H BuNo 137627 by AAA on 10 February 1966 during an attack on a bridge between Dong Hoi and the DMZ. Another bridge was dropped on the 10th by LT Kurt V. Anderson and his wingman who also destroyed four trucks and damaged seven others during the flight. On 25 April, a sick engine forced LTJG Malcolm Johns to make a forced landing at Duc Hoa AB. On the 26th, four aircraft destroyed 25 large buildings and 15 smaller ones at a VC storage complex southeast of Saigon. The strike was repeated on the 28th and 35 more buildings were destroyed and 40 more damaged. Tragedy struck the squadron again on 20 June when LCDR John W. Tunnell crashed after a night launch in BuNo 139806. On 3 August, CDR D.E. Sparks relieved CDR Griffith and a returning A-1H, BuNo 134586, with battle damage rolled off the side of the ship when its brakes failed. The pilot, LTJG David Franz, was recovered safely. VA-145 departed Yankee Station on the 5th

after spending 136 days on the line and expending five million pounds of ordnance during 1,700 combat missions.

The squadron's last Skyraider deployment was aboard the limited attack carrier USS Intrepid (CVS-11) from 11 May to 30 December 1967. Pre-deployment operations cost VA-145 two aircraft and one pilot. On 17 November 1966, LCDR Tinley L. Olton crashed into the sea off California in BuNo 139654 and died after striking the mast of a small boat. Then, on 23 January 1967, LCDR A.D. Windsor was forced to make a gear-up landing at Fallon, NV, after the engine failed in BuNo 134563.

CVS-11 arrived on Yankee Station on 19 June 1967 with only

Below VA-145 A-1H BuNo 134563 put down in a field due to a massive oil leak and subsequent engine failure near NAS Miramar, CA. Note that the whole belly is black from oil. (USN via Jim Sullivan)

eight Skyraiders and were mostly assigned RESCAP missions. The unit's first strike mission was conducted on 9 July when they first provided FAC over NVN artillery positions before attacking them themselves. On 28 July, CDR Walter J. Schutz relieved CDR Sparks and the RESCAP missions continued. During the squadron's 102 days on the line, the RESCAP missions resulted in the rescue of fourteen downed aviators.

Ordnance expended during the cruise was 45,950 rounds of 20mm ammunition, 192 5-inch rockets, 25,587 2.75-inch rockets and 300 250lb bombs. VA-145 left the war zone with two A-1Js and four A-1Hs.

Once back in CONUS, the last Skyraider was retired in January 1968 and the squadron was relocated to NAS Whidbey Island, WA, where VA-145 began its transition to the

Above, VA-145 A-1H BuNo 135338 in late 1967 while assigned to CVS-11. Fuselage stripe and tail tip were green. (Paul Minert collection) Bottom, VA-145 A-1J BuNo 142033 taxis past revetments in Vietnam in 1967 with an Air Force C-130 in the background. (via Tailhook)

Grumman A-6A Intruder.

VA-145

Above, VA-145 A-1J BuNo 142016 off Vietnam in 1967 while assigned to the USS Intrepid (CVS-11). (USN) At right, VA-145 A-1J BuNo 142016 taxis past revetments in Vietnam in 1967. (via Tailhook) Below, VA-145 A-1H BuNo 139820 off Vietnam in 1967. Note Tonkin Gulf Yacht Club patch on aft fuselage. (USN)

ATTACK SQUADRON SEVEN - TWENTY - EIGHT, VA-728
SECOND ATTACK SQUADRON ONE - HUNDRED - FIFTY - FIVE, VA-155

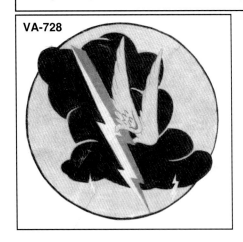

VA-728

Unofficial 1954-55 CVA-10 insignia.

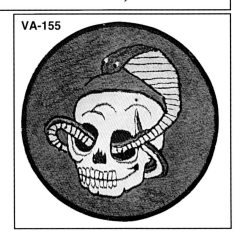

VA-155

VA-155 was originally established as Reserve Attack Squadron Seventy-One E (VA-71E) at NAS Glenview in 1946. VA-71E was redesignated Reserve Attack Squadron Fifty-Eight A (VA-58A) on 1 October 1040 and then Reserve Composite Squadron Seven Hundred Twenty-Two (VC-722) on 1 November 1949. And finally Reserve Attack Squadron Seven Hundred Twenty-Eight (VA-728) on 1 February 1951, when it was called to active duty in response to the Korean War. Up to 1951, as a reserve unit the squadron had flown Avengers, Helldivers and Maulers.

VA-728, while under the command of LCDR Soule T. Bitting, received its first Skyraiders, two AD-1s, in March 1951. They had twelve AD-1s by the end of April and report-

ed to NAS Alameda, CA, in June where they received twelve combat capable AD-4Ls. They retained the AD-1s until August to aid in training and on 9 September took nine AD-4Ls, seven AD-4s, and one AD-4Q aboard the USS Antietam (CV-36) for a Korean War deployment lasting until 2 May 1952.

CV-36 joined CV-9 and CV-31 in October 1951 and the three ships concentrated in rail cutting with 490 cuts being made between 18-31 October. VA-728s first operational loss occurred on 16 October when an AD-4L ditched astern of the ship. Two more AD-4s, BuNos 124002 and 124004, were lost after launch from CV-36 on 6 December. On 14 December, AD-4 BuNo 123825 was hit by AAA and ditched at sea. Then, on 21 January 1952, another aircraft,

AD-2 BuNo 122315, ditched at sea after being hit by AAA. During the cruise, ENS John Higgins took a round in the canopy narrowly missing his head and shattering the windscreen, which reduced his forward visibility to zero. Radio instructions from the LSO got him safely aboard.

Upon return to California, LCDR B.K. Harrison took command in May 1952 and the squadron operated out of Santa Rosa in June and July. In August, VA-728 moved to Moffett Field and on 8 October, LCDR R.S.

Below, bombed-up VA-728 Skyraider takes off from the USS Antietam (CV-36) in 1951-52 for a mission over Korea. Fin tip and prop hub were green. (National Archives)

Osterhoudt relieved CDR Harrison. On 24 January 1953, sixteen AD-4Ns were loaded aboard the USS Princeton (CV-37) for the squadron's second Korean War deployment. Eight days later, on 1 February 1953, VA-728 was augmented into the regular Navy and redesignated Attack Squadron One Hundred Fifty-Five (VA-155). During the cruise, interdiction was the order-of-the-day as the squadron concentrated on trains, rail lines, marshalling yards, bridges, and storage facilities with close air support missions thrown in when needed. Three aircraft were lost in July, the first, AD-4N BuNo 125757, crashed on 12 July 1953. The second, AD-4NA BuNo 126904, was shot down by AAA on 16 July and the third, AD-4N BuNo 126922, ditched off of CV-37 due to engine problems on 25 July. The deployment was concluded on 21 September 1953 when VA-155

VA-728 AD-4 operations aboard CV-36 off Korea. (NMNA)

returned to Moffett.

In October 1953, LCDR Frank R. West took command and in November the first nine AD-6s were acquired. The squadron then flew a mixed bag of AD-6s, AD-4Bs, AD-4NAs and AD-3s through April 1954. On 1 July 1954, VA-155 took sixteen AD-6s aboard the USS Yorktown (CVA-10) for a WestPac deployment which ended on 28 February 1955. After conducting operations in the South China Sea, VA-155 spent Christmas at Yokosuka. On 23 January 1955, they began 20-days of air patrols to protect the evacuation of the Tachen Islands by the Nationalist Chinese. During the cruise the unit was nicknamed the "Tigers".

Once back at Moffett, LCDR Jack B. Jones became CO on 31 March 1955 and an AD-5 was acquired to aid in training. CDR Jones took sixteen AD-6s aboard the USS Wasp (CVA-18) for a WestPac cruise from 23 April to 15 October 1956.

Back in CONUS, an AD-5 was acquired once again in October and

At left, two photos of VA-728 AD-4L BuNo 123968 hangs over the edge of the USS Antietam (CV-36) on 10 July 1951 after crashing into the barrier. (National Archives via Jim Sullivan) Below, VA-728 Skyraider BuNo 123999 H/512 shed its right main gear at Santa Rosa, CA, in June 1952. (USN)

258

in November the first five AD-7s were received. On 10 December 1956, CDR Henry E. Clark relieved CDR Jones and the squadron became known as "Baron Van Clark's Flying Circus". In October 1957, the last AD-6s were transferred out and a second AD-5 was added in December to complement the unit's twelve AD-7s. The squadron's last AD deployment was aboard the USS Hancock (CVA-

At top, VA-728 AD-4L tangled with the island on the USS Princeton (CVA-37) on 25 February 1953. (National Archives via Jim Sullivan) At right, VA-155 AD-6s line up to launch from CVA-10 in 1954. (USN) Below, bombed-up VA-728 Skyraider runs-up prior to taxi and deck launch from CVA-37 off the coast of Korea on 25 August 1953. (USN)

Above. VA-155 AD-6 BuNo 137525 with green trim on the tail and wing tips on 21 May 1955. (Jim Sullivan collection) At left, VA-155 AD-6 BuNo 137596 in flight with "Barn Door" open. Tail trim was green. (Paul Frieler) Below left, VA-155 AD-7 BuNo 142062 armed with 5" rockets and napalm taxis on CVA-19 in 1958. (USN) Bottom, VA-155 AD-7 BuNo 142043 takes off from the USS Hancock (CVA-19) in 1958 with rockets and napalm. (USN)

19) from 15 February to 2 October 1958. From 21 August to 11 September, VA-155 flew in support of the Nationalist Chinese once more, this time after the shelling of Quemoy Island.

Upon the squadron's return to the states, all the AD-7s were transferred out by the end of October 1958, but an AD-5 was retained during the transition to the A4D-2.

ATTACK SQUADRON EIGHT - FIFTY - NINE, VA-859
ATTACK SQUADRON EIGHTY - FIVE, VA-85 "BLACK FALCONS"

Reserve Attack Squadron Eight Hundred Fifty-Nine (VA-859) was established at NAS Niagara Falls, NY, and was called to active duty on 1 February 1951 in response to the Korean War. Commanded by LCDR Richard E. Moot, VA-859 was initially equipped with TBM-3E Avengers.

The squadron's first Skyraiders, four AD-2s, were received in March and the unit transferred to NAS Jacksonville, FL, on 5 April. They were equipped with seventeen AD-2s in May and by the end of June fielded one AD-4L and eighteen AD-2s. On 26 September, they moved to NAS Quonset Point, RI, where they began their final preparations prior to deploying on the USS Tarawa (CV-40). Carrier qualifications aboard CV-40 were conducted in October and November, before deploying to the Mediterranean from 28 November 1951 to 11 June 1952. The squadron visited Guantanamo Bay, France, Spain, Italy, Greece, and Turkey during the cruise.

Upon return from the Med, the squadron was relocated to NAS Oceana, VA. LCDR Joe W. Williams, Jr. assumed command on 26 September 1952 and oddly an F8F-2 and an F4U-4 were acquired. The F4U-4 transferred out in October and the F8F-2 left in January 1953. In February, the last AD-2 was retired and the squadron was equipped with four AD-4Ls, four AD-4Bs and ten AD-4s. On 4 February 1953, VA-859

was redesignated Attack Squadron Eighty-Five (VA-85). In March and April, carrier qualifications were conducted on the USS Coral Sea (CVA-43) in preparation for its Med deployment from 26 April to 21 October 1953. On 7 May, port call was made at Oran, Algeria, and on 24 June they became the first US warship to drop anchor in Barcelona, Spain. During the cruise, VA-85 took part in NATO exercise "Black Wave".

After returning to Oceana, VA-85 began transferring its mixed bag of AD-4s out in preparation for new AD-6s. One AD-6 was acquired in February 1954 prior to CDR Adolph Mencin assuming command in March. By May, VA-85 had a complete complement of eighteen AD-6s on-hand. In July, the unit boarded the

USS Lake Champlain (CVA-39) for refresher training at Guantanamo Bay, Cuba. The ship returned from Cuba in August and deployed to the Med in September, from 27 September 1954 to 15 April 1955. CVA-39 passed Gibraltar on 9 October and began operations with the 6th Fleet. During the cruise, fourteen port calls were made. These were: Leghorn, Rome, Genoa, Naples, Venice, Cannes, Marseille, Toulon, Salonika, Athens, Lebanon, Istanbul, Beirut, and Barcelona.

CDR Charles H. Jaep III

Below, VA-859 AD-2 BuNo 122234 after a barrier crash on the USS Tarawa on 7 July 1952. (National Archives)

261

assumed command on 6 May 1955 and an AD-5 was acquired that same month. In June, an AD-5N was added and refresher training was conducted aboard CVA-39. In November-December during the squadron's four week training period at Boca Chica Field, FL, the pilots flew 1,300 hours in just fifteen days. To accomplish this feat, a night check crew was run each night to insure maximum availability of the AD-6s each day. Training included long-range navigation, FCLP, and high altitude dive bombing. CarQuals in January-February 1956 were conducted aboard the USS Intrepid (CVA-11) in preparation for its 12 March to 5 September 1956 deployment to the Med. VA-85 completed the cruise with 3,818 flight hours and an aircraft availability of 86%.

Above, VA-85 AD-6 BuNo 135252 on display in 1954-55. Tail code was "E", fin tip green and edge of rudder was white. Note aircraft number 3 on the vertical fin. (Paul Minert collection) Below, VA-85 Skyraiders taxi for take-off aboard the USS Lake Champlain (CVA-39) on 6 November 1954 during Exercise Hellenic Sky 2. (USN)

CDR Jack C. Heishman became CO in September 1956. In early 1957, CarQuals were conducted aboard the USS Randolph (CVA-15) followed by night CarQuals aboard the USS Franklin D. Roosevelt (CVA-42). CDR M.G. Bramilla, Jr. assumed command in September 1957. VA-85 became the first squadron to report aboard the new supercarrier Ranger for its shakedown cruise to Guantanamo Bay from 7 October to 5 December 1957.

In March 1958, VA-85 conducted carrier qualifications aboard CVA-11 and in April and May was aboard the USS Forrestal (CVA-59) for a joint

Civilian Orientation cruise which ended with exercise LANTRAEX I-58. During the exercise, LTJGs Strang and Woods flew non-stop below 1,000 feet from Forrestal to North Island and back again to demonstrate the low-level, long-range ability of the AD. The flight took 10.5 hours coast-to-coast. CDR E.M. Coppola took command on 5 September 1958 and routine training activities continued. CDR Coppola was relieved by CDR Howard C. Lee on 25 March 1959.

During the last week in July 1959, VA-85 boarded Forrestal for six weeks of refresher training to Guantanamo Bay after the ship's

Above, VA-85 AD-6 BuNo 135225 at Bolling AFB on 22 May 1955. (NMNA) Bottom, VA-85 AD-6 BuNo 136405 in 1954. (NMNA)

overhaul. Back at Oceana, the squadron took first place in the prop light attack category at the East Coast Weapons Elimination Competition Meet which allowed them to attend the 4th Annual Naval Weapons Meet at Yuma from 30 November to 4 December. The meet was named Operation Top Gun and VA-85 won the "High Team" award in the VA/props category with CDR Lee

Above, VA-85 AD-6 BuNo 139642 taxis on CVA-15 on 22 March 1957 during carrier qualifications. (USN) Below, VA-85 AD-5 BuNo 134976 aboard the USS Intrepid (CVA-11) on 20 April 1956. Fin tip was red. (Naval History) Bottom right, VA-85 AD-6 BuNo 137570 at Yuma on 3 December 1959. (Harry Gann)

earning the "High Individual" award. Then VA-85 took its twelve AD-6s aboard CVA-59 for a Med cruise from 28 January to 31 August 1960.

After returning to CONUS, CDR William Carrier, Jr. took command on 29 November 1960 and by February 1961 VA-85 found itself on the way to the Med once again. They deployed on CVA-59 from 2 February to 25 August 1961. During the cruise, 5,100 hours were flown, a record for a deployed AD squadron.

CDR N.O. Scott, Jr. took command in December 1961 and in January 1962 was aboard CVA-59 for six weeks refresher training at Guantanamo Bay. They also participated in the recovery of Project Mercury's John Glenn and in April the Forrestal and VA-85 put on a firepower demonstration for President John F. Kennedy. The squadron's final A-1 deployment was aboard CVA-59 from

Above, VA-85 AD-6 BuNo 137569 at MCAAS Yuma, AZ, for the Naval Air Weapons Meet on 3 December 1959. USS Forrestal was in red and tail trim was in green. (Harry Gann) Below, four VA-85 pilots pose with their AD-6 aircraft on the USS Ranger to commemorate the first Skyraider landings on CVA-61. (Nat. Archives)

Above, VA-85 flight ops aboard CVA-59 on 1 March 1963 with A-1H BuNo 135323 (AJ/500) in the foreground and 135814 in the background. (USN) Below, VA-85 AD-6s 139607 (AJ/411) and 137571 (AJ/414) on the USS Ranger (CVA-61) in November 1957. (USN)

3 August 1962 to 2 March 1963. AD-5N BuNo 132672 was ditched safely off Forrestal on 18 October due to an engine failure. During the cruise, CDR Clinton H. Mundt relieved CDR Scott on 21 December 1962. Then, on 5 February 1963, CDR Mundt was killed when he crashed into the sea off the southern coast of Sicily. He was replaced by CDR John C. McKee who retained command until February 1964.

Back at Oceana, VA-85 continued to fly twelve A-1Hs and one A-1E until March

Above, VA-85 AD-6s from Forrestal bank over the Atlantic in 1960. (USN) Below, VA-85 AD-6 BuNo 139636 over the Med was from the USS Forrestal (CVA-59) on 18 July 1961. (USN)

1964 when it began receiving A-6A
Intruders. The last four A-1Hs were
retired in April.

**Above, VA-85 AD-6s BuNos 137532 (AJ/511) and 137503 (AJ/510) in flight over the
Mediterranean on 28 July 1960. (USN) Below, VA-85 AD-6 BuNo 139814 refuels a
Royal Navy Scimitar from HMS Hermes in November 1962. (USN) Bottom, four VA-
85 A-1Hs over the sea on 25 May 1959. (USN)**

ATTACK SQUADRON NINE - TWENTY - THREE, VA-923
FIRST ATTACK SQUADRON ONE-TWENTY-FIVE, VA-125 "ROUGH RAIDERS"

VA-923 patch at left. (D. Siegfreid)

VA-125 was originally established as Reserve Attack Squadron Fifty-Five E (VA-55E) at NAS St. Louis after WWII. Commanded by LCDR Herb W. Wiley and flying TBM Avengers, VA-55E was redesignated Reserve Attack Squadron Nine Hundred Twenty-Three (VA-923) on 1 January 1950. Most of the TBMs were replaced with AM-1 Maulers before the squadron was called to active duty on 20 July 1950. On 2 August, VA-923 transferred to NAS San Diego, CA, where they joined CVG-102. Eleven AD-2s were received in August and another four in September along with an AD-4Q. The AD-2s were replaced slowly and by the end of January 1951, there were three AD-4s, one AD-4Q, eleven AD-3s, one AD-2, and five AD-1s on hand.

VA-923 deployed to Korea aboard the USS Bon Homme Richard (CV-31) from 10 May to 17 December 1951 with fifteen AD-3s and three AD-4Qs. The squadron's first combat operations began on 31 May in support of UN troops north of Hwachow. This was followed by seventeen days of interdiction against rail and highway bridges, truck convoys, troop positions and warehouses and barracks. After rest and replenishment in Japan, combat operations resumed from 1-27 July. On 24 August, in coordination with Air Force bombers, VA-923 hit targets at Rashin and Najin and bridges north of Chongjin. On 9 October, ADs with 2,000 lb bombs and F4Us attacked the mining center

35 miles from Songjin, destroying the ore plant and destroying or damaging thirty-one buildings. On 27 November, several MiG-15s jumped the Air Group's ADs and F4Us with one AD being slightly damaged.

Below, bombed-up VA-923 Skyraider deck launches from the USS Bon Homme Richard in 1951 for a strike over Korea. (USN) Bottom, VA-923 flightline during the Korean War at San Diego. (Fred Roos collection)

Nine Skyraiders were lost during the cruise: AD-3, BuNo 122746, crashed and burned during an emergency landing at Kangnung on 6 July; Engine failure caused the ditching of AD-3, BuNo 122768, after launch from CV-31 on 7 July; AD-3, BuNo 122760, was shot down by AAA over Korea on 18 July; AD-3, BuNo 122730, was hit by ground fire on 27 September and was safely ditched at sea; AD-3, BuNo 122753, was hit by AAA and ditched at sea on 3 October; AD-3, BuNo 122852, was ditched at sea after being hit by AAA on 6 October; AD-2, BuNo 122346, was hit by AAA south of Wonsan on 4 November with the pilot bailing out successfully; AD-3, BuNo 122767, was hit by AAA over Korea and the

pilot was lost on 21 November; AD-3, BuNo 122782, disappeared near Hwansungwon-ni in December 1951.

After returning to San Diego and being reassigned to NAS Miramar, CDR John C. Micheel took command on 24 March 1952 and prepared the squadron for its next war cruise. Three two-week weapons tours at NAAS El Centro were conducted during which LT Joe House was killed during a midair with a trailing jet. During late August and early September, CarQuals were conducted on the USS Oriskany (CV/CVA-34) followed by the squadron's deployment from 15 September 1952 to 18 May 1953. CVA-34 arrived on line on 31 October and began combat

Above, VA-923 AD-3 extending wings prior to a mission over Korea from CV-31 in 1952. Offensive load is twelve 250lb bombs and three napalm tanks. (NMNA) Bottom, VA-923 AD-3 BuNo 122737 traps aboard CV-31 off Korea in 1951. (NMNA)

operations on 2 November. On the 4th, ENS Andy Riker was shot down and killed by AAA southwest of Wonsan in AD-3 BuNo 122823. Then on the 15th, LT George Gaudette was killed in the crash of BuNo 122786. On 4 December, LTJG Jim Hudson was killed when his AD-4, BuNo 127862, was shot down by AAA near Pyongyang. The squadron flew com-

bat during five line periods: 31 October to 22 November 1952, 2 to 27 December 1952, 7 January to 13 February 1953, 1 March to 1 April 1953, and 7 April to 28 April 1953. VA-923's score for the first three line periods was the expenditure of 1,363 tons of bombs and rockets and of 44,374 rounds of 20mm. Targets destroyed or damaged were:

TARGET	DESTROYED	DAMAGED
Structures	62	99
Bunkers	38	46
Bridges		7
Vehicles	2	4
Rail Cars	12	36
Rail Cuts		36
Roundhouse	1	1
Turntables	1	
Railroad Tunnels		3
Caves		9
Road Cuts		20
Dock Facilities		1
Fuel Facilities	4	1
Gun Emplacements	14	25
Radar	1	4
Storage Shelters	176	183
Supply Tunnels		1
Supply Stocks	35	5
Electrical Transformers	1	

Three days before VA-923 was redesignated VA-125 on 4 February 1953, CO CDR Micheel was shot down and killed in BuNo 122822 while leading a bridge strike. The XO, LCDR A.H. Gunderson, assumed command that same day on 1 February.

After the squadron's return to Miramar, there was a shortage of pilots and planes for the first couple

Above, VA-923 AD-3 and VF-874 F4U-4s waiting to deck launch from CVA-31. (USN) Below, Harold Reutebuch waves through shell shatered canopy in his VA-923 AD-4 aboard CV-31 on 8 July 1951. (Jim Sullivan collection) Bottom, VA-923 AD-3 BuNo 122737 "Hefty Betty" runs-up prior to deck launch from CV-31 in October 1951. (John Wood via Fred Roos/J. Sullivan)

months and in July 1953 when LCDR John L. McMahon, Jr. assumed command, there were five AD-4Bs, ten AD-3s, and one AD-1 on-hand. On 3 March 1954, VA-125 took eight AD-4Bs and eight AD-4NAs aboard the USS Boxer (CVA-21) for a WestPac cruise. The deployment ended on 11 October 1954.

The squadron's fourth deployment was aboard the USS Hancock (CVA-19) from 10 August 1955 to 15 March 1956 whlie under the command LCDR Bernard H. Bahlman.

For VA-125's last deployment, CDR Bahlman remained in command. The WestPac deployment was aboard the USS Lexington (CVA-16) from 19 April to 17 October 1957. Prior to deployment, 41 Navy "E"s were awarded the unit's pilots for proficiency in special weapons, strafing, rocketry, dive bombing, night bombing and loft bombing.

After the cruise, the squadron returned to Miramar where CDR A.J. Henry Jr. took command on 25 October 1957. VA-125 was disestablished on 10 April 1958.